D0208403

HOW TO

CALCULATE

QUICKLY

(the art of calculation)

BY HENRY STICKER

DOVER PUBLICATIONS, INC.

This Dover edition, first published in 1955, is an
unabridged republication, with minor corrections,
of the work originally published by Essential Books
in 1945 under the title *The Art of Calculation*. It
is reprinted through special arrangement with
Duell, Sloan and Pearce, Inc.

International Standard Book Number: 0-486-20295-X
Library of Congress Catalog Card Number: 56-3700

. Manufactured in the United States of America
Dover Publications, Inc.
31 East 2nd Street, Mineola, N.Y. 11501

PREFACE

Arithmetic is a science, but calculation is an art. Science is knowledge—art is skill. You have all the knowledge you could possibly need to determine that 57 times 25 equals 1425, but if you are asked to multiply 57 by 25 and cannot do this mentally in just about one second, you are not adept at the art of calculation.

Genuine skill in the calculating art can be acquired by any person of ordinary intelligence, no matter what his schooling may have been. To develop such skill is the purpose of this book. Special forms of short, graded exercises, performed for the most part mentally, lead the student by easy steps to a point where he will possess really exceptional calculating ability.

For instance, if you will look at Exercise No. 371 on page 191, you will find that you are expected to perform mentally such multiplications as 696 times 858, 858 times 878, etc. These are not "trick" examples—the student who systematically performs the practice examples presented in this book will be able to do many kinds of examples of this degree of difficulty by his sheer ability to hold and manipulate figures *in his head*.

How is this skill developed? Essentially by developing *number sense*. Number sense consists in the ability to recognize the relations that exist between numbers considered as whole quantities, and to work with the thought of their broad relations always uppermost. Number sense is possessed by many people in all walks of life—particularly by accountants, bookkeepers, estimators, cashiers, storekeepers and the like. On the other hand, it is absent in many who have an excellent understanding of advanced mathe-

matics. The engineering professions are full of
those who require slide rules to perform calculations which the average billing clerk would do
mentally.

To give an example of what is meant by number sense, suppose you were asked to multiply
mentally 11625 by 12. If you felt at all compe-
tent to try to do so, you would probably (unless
you are the exceptional case) proceed like this:
12 times 5 is 60, remember 0 and carry 6; 12
times 2 is 24, put 0 before the other 0 and carry
3, etc. In this way you would eventually arrive
at the correct answer—if you did not get all
mixed up in the meantime; but at best you would
take a long time, because number sense would
have played no part whatever in your awkward
method of approaching this very simple little
problem.

Suppose now that we introduce a little of this
number sense—suppose that instead of dealing
with plain figures, you were told to imagine that
you had sold twelve machines on each of which
you made a commission of \$11.62½. As soon as
money enters into the matter you immediately
see the whole picture in a different light. If you
were asked *approximately* how much your com-
missions amounted to, you would figure quick as
a flash that 11 times 12 is 132, and you would
probably answer instantly that you had made
something over \$132. If you were then asked
how much over \$132, you would either figure that
62½¢ are ⅝ of one dollar, or else that this amount
is equal to half a dollar plus ⅛ of a dollar. You
would not take long in determining that the excess over \$132 comes to \$7½, and that therefore the

total amount received would be \$139½ or \$139.50.
Why not apply to numbers "in the raw" the
same methods that you use when dealing with
small amounts of dollars and cents? It is no more
difficult to multiply 11⅝ thousands by 12 than
11⅝ dollars. If 11⅝ dollars times 12 is 139½ dollars, then 11⅝ thousands times 12 is 139½ thousands, or 139,500.

From this illustration you may correctly infer
that the person with number sense works very
largely *from left to right* instead of from right to
left. Left-to-right calculation is of the essence of
number sense. Countless practical people know
this, yet the art of left-to-right calculation is
never taught in the schools, and is, in fact, rarely
mentioned in books of any kind.

Step-by-step instruction and practice in this
neglected art of left-to-right calculation constitutes the greater part of the substance of this
book. Methods of this kind are applied not only
to multiplication but to all the fundamental operations. By means of such methods, for instance, you learn to add two columns of figures
at a time, and you even get a little practice in
three-column addition. You are also taught
comparable methods of subtraction and division.

In addition to the exercises having to do with
left-to-right calculation, there are many that are
based on an *extension of the multiplication table*.
You are taught by easy stages to use all the numbers up to 25 as direct multipliers—that is to say,
you acquire a *complete* knowledge of the multiplication table up to 25 times 25.

The subject of fractions is treated with special
reference to the addition and subtraction of the

fractions that are most commonly met with in everyday work. The object here is to enable the student to memorize the answers to the kinds of problems that are ordinarily figured out over and over again.

The exercises dealing with decimals are designed to give the student a large workable fund of knowledge of the decimal equivalents of fractions. Memory work includes twelfths and sixteenths, and there is practice in the rapid calculation of thirty-seconds and twenty-fourths.

The final broad subject developed in this book is "short cuts." These are of the highest value in developing a general understanding of numbers.

The subject matter of this book is limited to the four fundamental operations, with the inclusion of fractions and decimals. No attempt is made to consider the various fields of arithmetical application. Skill in calculation pure and simple is the only goal.

The exercises, nearly four hundred in number, are for the most part very short. Few should take more than ten minutes to do, and many will take less. As progress is by graded steps, the instruction is in small "doses." The book, accordingly, can be used with profit whenever you happen to have a few free minutes. Its pocket size, moreover, makes it all the more suitable for odd-moment study.

Taken as a whole, this book will prove valuable to anybody engaged in work or study that requires any considerable amount of arithmetical calculation. It is especially recommended to heads of departments in industrial and commercial organizations, for general distribution to the members of their staffs.

CONTENTS

1

THE PLAN OF THIS BOOK

The subject matter here presented might have been divided into sections on addition, subtraction, multiplication, etc., in the manner usual to text-books on arithmetic. Because, however, of the special purpose of this book, no such division is made. The general plan is to have several branches proceed simultaneously. Progress is not from subject to subject but from less to more difficult calculation.

For each of the fundamental divisions of arithmetic there is a general introduction—for instance, *Addition in General* on page 3 . In these introductions the special objects sought are described, as well as the methods by which these objects are attained. The student, therefore, always has a clear view of the ultimate aims of his studies and knows how the work immediately in hand fits into the general plan.

Wherever anything new is introduced, it is clearly explained and illustrated. Usually the exercises that go with each explanation are spread through many succeeding pages. In a large number of cases the exercise calls for work with the numbers in a certain list or table (for instance, Table I on page 7). The same lists of numbers are used for various kinds of calculation. This method of presentation makes possible the remarkably great number (about 15,000) of practice examples that are included.

ADDITION IN GENERAL

Two main objects are sought. The first is to add by single columns, grouping three successive numbers at a time; the second is to add two columns at a time:

Take the following sum:

$$26$$
$$43$$
$$84$$
$$72$$
$$96$$
$$27$$
$$42$$
$$35$$
$$68$$
$$64$$
$$37$$
$$97$$

By the first method, starting at the top of the units' column, we would add these numbers thus: (sum of the first three figures) 13 (+ sum of the next three figures, 15) 28 (+ 15) 43 (+ 18) 61; write 1 and carry 6; (6 + 14) 20 (+ 18) 38 (+ 13) 51 (+ 18) 69; total, 691.

By the second method, starting at the top, we would add both columns simultaneously thus: (26 + 43) 69 (+ 84) 153 (+ 72) 225 (+ 96) 321 (+ 27) 348 (+ 42) 390 (+ 35) 425 (+ 68) 493 (+ 64) 557 (+ 37) 594 (+ 97) 691.

In actual practice, very rapid addition is possible by either method, and you will be left free to choose whichever you prefer. You should, however, learn both.

How do you proceed to learn these methods? You were taught—or should have been taught—at school that speed in addition is acquired by combining pairs of successive numbers that add up to 10. It is at this point that we start, because this is the simplest way in which grouped numbers can be added to a preceding sum. You are given short columns of numbers to be added by incidentally selecting such pairs of successive figures as make 10. In succeeding exercises the columns are lengthened, and you are also asked to group any pairs that add up to less than 10.

In the meantime, you will have been doing exercises in mentally adding all the numbers from 11 to 18 to all the numbers from 1 to 99. Since no pair of figures in a column can add to more than 18, this amount of practice will enable you to add *any* pair of successive figures in a column to a previous sum, and hence to add the entire column by taking two figures at a time.

You are similarly taught to add trios of numbers that make 10 or less than 10, and to add any number from 19 to 27 to any number from 1 to 99. With this practice you will be able to add *any* column by taking three figures at a time.

If you can quickly add any number from 1 to 27 to another number, you will not find it difficult to add numbers greater than 27 in the same manner. You are accordingly ready now to add two columns at a time. Exercises in this method are introduced, and these are gradually increased in difficulty.

Toward the end of the book there are some exercises in three-column addition—just enough to demonstrate that it will be possible for *you* to add this way if you wish to use this method.

There are examples in addition of still another kind. These are not included for practice in addition as such but have a special bearing on the art of multiplying mentally. We need not consider sums of this kind at this point.

You will note that in the exercises in one-column addition you are alternately instructed *to add from the top down* and *to add from the bottom up*. In practical work it is of course immaterial in which direction addition is performed. You should, however, be able to add with equal facility in either direction, and by alternating as suggested you will get the necessary practice.

Exercise No. 1

Pairs Adding to 10

Add the following columns by grouping pairs of numbers that make 10. *Add from the top down.*

Thus you would add the first column by saying to yourself: 7, 17, 22, 32.

Do not consciously repeat in your mind anything but the successive totals. That is to say, do *not* add this column thus: 7 + 10, 17, +5, 22, +10, 32.

For another illustration of the correct method, take the second example. This is added thus: 8, 18, 20, 30.

Write your answers in succession on a piece of paper and compare them with the correct answers on page 154. (A good plan is to place the edge of your paper immediately under the examples, write the answers along this edge, and fold it under as it becomes used up.)

1. 7	**2.** 8	**3.** 4	**4.** 5	**5.** 6	**6.** 5
6	9	5	2	4	5
4	1	5	8	6	3
5	2	5	4	3	6
1	3	4	1	2	4
9	7	6	9	8	8

7. 5	**8.** 3	**9.** 8	**10.** 6	**11.** 5	**12.** 9
4	2	2	9	5	6
6	7	9	1	3	4
6	3	8	5	2	8
3	1	1	4	4	1
7	2	9	6	6	7

13. 3	**14.** 1	**15.** 6	**16.** 6	**17.** 1	**18.** 7
7	9	4	3	3	6
6	9	4	7	7	2
2	1	5	2	9	8
8	5	4	2	3	5
8	4	3	5	7	5

19. 1	**20.** 1	**21.** 6	**22.** 3	**23.** 7	**24.** 4
9	5	4	4	5	9
4	5	7	6	5	1
3	9	6	4	3	3
9	4	3	6	6	2
1	6	7	3	2	8

Table I

Numbers from 1 to 99

1	8	15	22	29	36	43	50
57	64	71	78	85	92	99	6
13	20	27	34	41	48	55	62
69	76	83	90	97	4	11	18
25	32	39	46	53	60	67	74
81	88	95	2	9	16	23	30
37	44	51	58	65	72	79	86
93	7	14	21	28	35	42	49
56	63	70	77	84	91	98	5
12	19	26	33	40	47	54	61
68	75	82	89	96	3	10	17
24	31	38	45	52	59	66	73
80	87	94					

Exercise No. 2

Mental Addition

Add 11 to each of the numbers in Table I above.

Use *left-to-right* addition, which is performed by first adding the tens of one number to the whole of another. In other words, starting with the number in the table you first add 10 and then 1. A few illustrations will be in order:

15 + 11: say 15, 25, 26;

22 + 11: say 22, 32, 33;

29 + 11: say 29, 39, 40;

99 + 11: say 99, 109, 110.

Work down the columns—not across the page. Write down your answers and compare them with those on page 154.

Exercise No. 3
Pairs Adding to 10

Group all pairs of successive numbers that make 10.
Add from the bottom up.

1. 7	2. 6	3. 5	4. 9	5. 6	6. 3
8	4	2	7	7	1
4	5	5	6	9	6
6	2	4	4	1	4
5	4	6	8	3	4
3	5	6	8	4	1
5	4	7	9	6	8
5	1	3	1	3	2
1	2	4	1	8	9
8	8	8	7	5	6
2	7	2	5	2	4
<u>5</u>	<u>3</u>	<u>4</u>	<u>5</u>	<u>8</u>	<u>7</u>

7. 4	8. 8	9. 4	10. 6	11. 9	12. 3
7	2	4	5	8	7
3	9	3	7	8	6
8	1	2	3	2	6
3	5	4	4	7	1
2	3	6	2	1	2
2	8	1	8	9	7
8	5	6	9	6	6
1	5	4	1	5	4
9	2	9	3	5	5
1	6	3	2	5	5
<u>9</u>	<u>5</u>	<u>7</u>	<u>1</u>	<u>4</u>	<u>6</u>

13. 7	14. 3	15. 9	16. 1	17. 3	18. 6
4	7	1	8	6	9
6	8	6	7	4	1
3	2	3	5	2	7
2	8	7	5	8	7
6	5	5	6	5	3
4	5	4	7	1	2
1	8	6	3	4	1
8	2	4	5	1	5
3	7	3	4	9	2
7	1	2	4	3	9
9	9	9	6	7	1

Exercise No. 4
Mental Addition

Add 12 to the numbers in Table I on page 7.
To illustrate:

15 + 12: say 15, 25, 27;
22 + 12: say 22, 32, 34;
29 + 12: say 29, 39, 41;
99 + 12: say 99, 109, 111.

Exercise No. 5
Mental Addition

Add 13 to the numbers in Table I on page 7.

Exercise No. 6
Mental Addition

Add 14 to the numbers in Table I on page 7.

Exercise No. 7

Mental Addition

Add 15 to the numbers in Table I on page 7.

Exercise No. 8

Pairs Adding to 10 or Less

The grouping of pairs of successive numbers is now to be extended to include any that add to less than 10 as well as any that add to 10. That is to say, as you add each column watch to see whether any two successive numbers add to either 10 or less than 10, and if they do, make one addition of them to the preceding sum.

For this exercise use the columns of numbers in Exercise No. 1 and compare your answers with those for Exercise No. 1. *Add from the top down.*

To illustrate, the first column is added: 7, 17, 23, 32; the second: 8, 18, 23, 30; the third: 9, 19, 29.

Exercise No. 9

Mental Addition

Add 16 to each of the numbers in Table I on page 7.

Exercise No. 10

Mental Addition

Add 17 to each of the numbers in Table I on page 7.

Exercise No. 11
Pairs Adding to 10 or Less
Add the columns in Exercise No. 3 by grouping all pairs of successive numbers that add to 10 or less than 10. *Add from the bottom up.*

Exercise No. 12
Mental Addition
Add 18 to each of the numbers in Table I on page 7 .

Exercise No. 13
Adding Single Columns by Pairs
Add the following by single columns, taking pairs of successive numbers at a time. *Add from the top down.* The first example would be added: 5, 14, 25, write 5 and carry 2; 2, 12, 27, 36; answer 365.

1. 43	2. 29	3. 58	4. 87	5. 16
62	75	33	62	91
78	36	65	94	33
81	69	98	27	56
14	43	72	89	29
87	16	45	74	32

6. 19	7. 48	8. 77	9. 36	10. 63
99	21	29	49	78
36	68	49	94	96
71	29	11	59	44
61	18	51	22	41
41	25	53	27	88

11. 33	12. 21	13. 34	14. 24	15. 16
39	79	43	14	44
43	74	27	11	49
51	85	53	15	54
55	63	17	75	49
36	82	57	78	99

16. 31	17. 28	18. 63	19. 32	20. 63
35	63	35	65	28
67	21	12	16	76
44	34	31	67	45
84	52	81	73	69
42	56	15	55	62

21. 85	22. 54	23. 14	24. 68	25. 69
56	42	27	42	28
75	68	54	28	45
37	13	85	34	37
73	99	59	83	71
24	84	69	16	91

Exercise No. 14
Mental Addition
Add 19 to each of the numbers in Table I on page 7.

Exercise No. 15
Adding Single Columns by Pairs

Add the following by single columns, taking pairs of successive numbers at a time. *Add from the bottom up.* The first example would be added: 11, 15, 27, 42, 49, 60, write 0 and carry 6; 6, 17, 24, 37, 43, 54, 62; answer, 620.

1. 27	2. 81	3. 92	4. 16	5. 29
64	28	92	14	27
32	75	29	14	25
85	43	86	31	25
46	96	54	97	32
29	57	18	65	19
78	51	68	29	76
64	89	62	79	51
31	75	11	73	12
43	42	86	22	84
75	54	53	58	33
46	86	65	64	19

6. 43	**7.** 58	**8.** 74	**9.** 91	**10.** 99
51	54	69	85	13
38	62	65	91	96
36	49	74	76	13
37	47	71	85	87
33	36	58	82	96
41	34	47	69	93
87	52	35	58	87
62	98	63	37	69
23	73	31	74	47
95	34	84	42	75
44	27	45	95	53

11. 19	**12.** 39	**13.** 51	**14.** 63	**15.** 84
12	41	55	62	99
26	23	52	62	75
18	37	34	63	73
24	29	48	45	74
24	35	56	59	56
18	98	46	67	82
15	29	31	57	78
98	26	53	42	68
36	91	37	64	53
85	48	13	48	59
49	96	59	24	57

Exercise No. 16
Mental Addition
Add 20 to each of the numbers in Table I on page 7.

Exercise No. 17
Adding Single Columns by Pairs
Add the following by single columns, taking pairs of successive numbers at a time. *Add from the top down.*

1. 51	2. 42	3. 41	4. 34	5. 33
30	53	73	36	81
96	90	32	97	28
24	79	12	19	39
25	87	62	69	43
75	76	11	94	10
48	92	44	83	85
49	52	84	37	47
93	45	70	38	29
80	72	40	46	14
13	18	61	17	95
58	63	67	23	10
88	22	56	66	82
86	21	16	64	31
20	59	98	89	77
99	91	55	68	74
59	15	27	60	35
65	78	54	23	84

6. 61	7. 34	8. 39	9. 36	10. 17
81	90	32	25	66
82	86	21	97	28
24	85	49	96	74
59	16	87	52	84
95	58	33	30	15
53	64	48	63	67
37	47	11	94	93
27	23	60	35	73
31	45	20	62	69
92	44	70	51	10
83	65	26	91	29
80	72	55	88	79
38	68	57	43	78
54	42	12	19	22
98	40	46	14	13
41	89	75	56	76
77	99	18	42	39

Exercise No. 18

Mental Addition

Add 21 to each of the numbers in Table I on page 7.

SUBTRACTION IN GENERAL

In keeping with the general object of this book, the succeeding exercises in subtraction are performed by left-to-right methods. When subtraction is performed on paper there is no special advantage in working from left to right instead of from right to left. Paper practice in the former method, however, fits in with the broad purpose of developing number sense. When it comes to doing subtraction mentally, the left-to-right method is natural and logical. Thus, if you had started the day with $17.43 in your pocket, and if you wanted to figure without paper and pencil how much you had left after spending $5.89, you would not be likely to start by subtracting 9 from 13. You would probably calculate that if you had spent the full $6, you would have $11.43 left, but that having spent 11¢ less than $6, the remainder comes to 11¢ more than $11.43, or $11.54.

In considering the specific aims of these exercises in subtraction, look first at the written examples. If you will glance at the first exercise that follows, and which is included merely to familiarize you with the idea of working from left to right, you will see that in every case the figures in the subtrahend (lower number) are smaller than those in the minuend. The examples are all of the type of

$$
\begin{array}{r}
54 \\
-23 \\
\hline
\end{array}
$$

and you can determine the answers faster than you can write them down. If, however, you take the example

$$685$$
$$-356$$

and try to write the answer in the same way, you will run into trouble when you reach the final figures at the right because 6 is greater than 5. What to do about cases of this kind is the subject of the instruction. The exercises take into account the possible variations that may occur in numbers of two and three places.

The examples in mental subtraction are performed by methods altogether different from those that apply to written work. There are two such methods, of which one has already been illustrated. We subtracted $5.89 from $17.43 by taking $6 from $17.43 and then adding to $11.43 the difference between $6 and $5.89, obtaining as our answer $11.43 + $.11, or $11.54. To do the same example mentally by the other method, we calculate that if you had started with $17 even, you would have $11.11 left; but you had $.43 more than $17 at the start, and therefore the actual remainder is $11.11 + $.43, or $11.54. One method is as good as the other. Examples are given that carry the practice in both methods as far as numbers involving hundreds of dollars and odd cents.

Incidentally, you should know that ordinary written subtraction is commonly performed by two entirely different methods—the *borrow*

method and the *carry* method. The borrow
method is taught almost exclusively in this coun-
try today, but in times past the carry method
had similar acceptance.

Take the example

$$
\begin{array}{r}
856 \\
-569 \\
\hline
287
\end{array}
$$

To do this by the borrow method you reason: 9
from 16 leaves 7, 6 from 14 leaves 8, 5 from 7
leaves 2; answer, 287. To do the same example
by the carry method you would say: 9 from 16
leaves 7, 7 from 15 leaves 8, 6 from 8 leaves 2;
answer, 287.

You should understand both these methods
(neither of which has any clear advantage over
the other), though you continue to use regularly
whichever one comes most naturally to you. In
the illustrations given in this book the borrow
method is followed because it is the more familiar
to the majority of people.

Exercise No. 19
Left-to-Right Subtraction

Perform the following subtractions by directly writing
your answers from left to right.

1. 67	2. 48	3. 41	4. 78	5. 64
55	14	20	22	31

6. 98	7. 53	8. 65	9. 28	10. 66
20	41	52	16	45

11. 99	12. 69	13. 83	14. 32	15. 93
92	35	31	21	41

Exercise No. 20

Left-to-Right Subtraction

Directly write your answers from left to right.

To take the first example, you simply note that 6 is greater than 4, and therefore the 5 in the minuend becomes a 4: 2 from 4 leaves 2 (writing 2), 6 from 14 leaves 8 (writing 8); answer 28.

1. 54	2. 47	3. 51	4. 46	5. 52
26	19	39	27	37

6. 84	7. 37	8. 35	9. 72	10. 50
58	18	17	24	29

11. 83	12. 56	13. 71	14. 96	15. 77
44	39	45	38	49

16. 94	17. 45	18. 48	19. 68	20. 71
76	16	29	39	52

Exercise No. 21

Mental Addition

Add 22 to each of the numbers in Table I on page 7.

Exercise No. 22

Trios that Add to 10 or Less

This exercise introduces the idea of taking in three successive numbers at a glance. Every column contains four groups of three numbers each; each of these groups adds to 10 or less. Add by combining these groups. *Add from the top down.*

1. 27	2. 14	3. 64	4. 57	5. 34
21	11	21	31	31
11	12	13	12	11
45	33	44	56	54
41	21	42	21	42
13	13	22	23	13
65	25	43	56	52
12	21	32	12	31
12	24	33	12	22
25	35	78	45	44
11	12	11	21	31
11	13	11	12	14

6. 41	7. 62	8. 43	9. 21	10. 33
21	32	33	11	12
26	12	24	15	15
31	61	21	12	63
31	21	11	11	11
22	23	27	14	24
81	52	43	33	42
11	21	11	11	22
11	16	45	23	44
72	44	62	24	43
21	12	12	21	32
13	14	15	25	33

Exercise No. 23

Left-to-Right Subtraction

Sight practice with pairs of three-place numbers. No borrowings are involved. Work from left to right.

1. 754	2. 827	3. 468	4. 659	5. 746
233	614	235	338	415

6. 928	7. 675	8. 558	9. 649	10. 458
615	423	146	437	328

11. 727	12. 898	13. 753	14. 462	15. 941
605	457	321	111	720

Exercise No. 24

Mental Addition

Add 23 to each of the numbers in Table I on page 7.

Exercise No. 25

Mental Addition

Add 24 to each of the numbers in Table I on page 7.

Exercise No. 26
Adding Single Columns by Pairs
Take successive pairs at a time. *Add from the top down.*

1. $40.72	2. $35.51	3. $27.13	4. $47.15
33.32	56.28	96.92	10.20
98.21	43.90	22.07	36.09
29.05	49.44	38.71	59.73
53.69	84.57	58.94	55.70
79.66	99.61	34.88	85.54
83.97	24.25	60.26	31.78
45.77	16.23	65.14	11.12
42.63	80.17	18.19	52.48
46.68	82.67	89.30	87.81
64.39	86.93	41.75	74.01
37.62	91.76	50.95	25.60

5. $79.45	6. $77.52	7. $48.68	8. $88.09
85.30	54.05	49.99	44.80
70.46	61.65	14.78	75.03
83.73	76.29	11.12	36.53
69.97	74.43	90.55	95.96
34.21	38.10	17.18	62.39
64.81	87.37	15.50	82.01
20.72	63.25	56.47	26.13
60.26	32.93	67.06	33.28
31.57	22.98	19.16	42.71
59.86	89.84	41.40	94.66
58.35	91.23	56.15	10.34

Exercise No. 27

Left-to-Right Subtraction

In these examples, in the vertical pairs of figures at the extreme right the subtrahend is greater than the minuend, reducing by 1 the tens' figure of the minuend.

Taking the first example, we note that the tens' figure of the minuend will become a 4 instead of a 5; 5 from 7 leaves 2, 3 from 4 leaves 1, 9 from 14 leaves 5; answer 215.

1. 754	2. 863	3. 528	4. 642	5. 995
539	448	319	313	217

6. 422	7. 323	8. 676	9. 266	10. 583
313	109	428	138	346

11. 912	12. 365	13. 744	14. 390	15. 555
509	259	619	265	419

16. 983	17. 696	18. 472	19. 713	20. 626
779	587	329	606	318

21. 718	22. 683	23. 951	24. 648	25. 873
409	246	229	539	358

26. 715	27. 582	28. 246	29. 997	30. 737
506	246	139	129	318

Exercise No. 28

Mental Addition

Add 25 to each of the numbers in Table I on page 7.

Exercise No. 29
Mental Addition
Add 26 to each of the numbers in Table I on page 7.

Exercise No. 30
Mental Addition
Add 27 to each of the numbers in Table I on page 7.

Exercise No. 31
Trios that Add to 20 or Less
In the separate columns of the following examples the successive groups of three figures add to some number between 11 and 20. Add by combining these groups of three. *Add from the top down.*

The first example would be added: 16, 30, 41, 61, write 1 and carry 6; 6, 18, 30, 46, 62; answer 621.

1. 23	2. 31	3. 12	4. 24	5. 24
46	46	84	64	74
67	46	89	74	78
21	12	33	35	35
55	24	43	45	55
58	97	78	95	78
22	13	13	14	14
54	73	37	45	44
95	86	99	75	99
12	23	13	25	25
69	57	88	65	35
99	77	98	86	69

6. 33	7. 32	8. 24	9. 34	10. 24
36	44	67	54	75
98	58	69	56	85
11	13	36	25	35
25	33	47	25	56
89	77	87	89	86
13	23	13	24	14
77	57	48	64	55
75	88	69	97	56
23	31	14	35	25
56	46	99	55	36
69	68	98	67	77

Exercise No. 32
Left-to-Right Subtraction

In the type of example given here we see by inspection that the subtrahend has a larger figure than the minuend in the tens' place, reducing by 1 the hundreds' figure of the minuend. To take the first example: 5 from 6 leaves 1, 9 from 15 leaves 6, 3 from 4 leaves 1; answer 161.

Subtract from left to right.

1. 754	2. 648	3. 262	4. 548	5. 629
593	356	191	357	458

6. 856	7. 435	8. 468	9. 914	10. 765
792	183	271	291	481

11. 787	**12.** 547	**13.** 341	**14.** 112	**15.** 783
693	160	171	51	190

16. 486	**17.** 888	**18.** 489	**19.** 944	**20.** 842
291	494	194	452	161

Exercise No. 33

Left-to-Right Subtraction

In these examples the tens and the units are larger in the subtrahend than in the minuend, thus reducing by 1 both the hundreds and the tens of the minuend. Taking the first example: 2 from 6 leaves 4, 8 from 14 leaves 6, 9 from 14 leaves 5; answer, 465.

1. 754	**2.** 773	**3.** 413	**4.** 484	**5.** 342
289	194	249	298	189

6. 626	**7.** 787	**8.** 383	**9.** 867	**10.** 672
578	298	197	379	295

11. 918	**12.** 666	**13.** 586	**14.** 232	**15.** 515
589	197	298	176	299

16. 353	**17.** 428	**18.** 856	**19.** 481	**20.** 318
169	179	779	192	149

Exercise No. 34
Adding Single Columns by Pairs

Add the following by single columns, taking pairs of successive numbers at a time. *Add from the bottom up.*

1.	2.	3.	4.
$14.44	$80.54	$74.43	$43.93
38.42	33.20	67.27	32.06
72.09	13.40	18.02	94.34
61.90	55.95	21.60	97.86
63.26	10.17	25.98	30.29
56.78	75.79	96.45	36.47
73.76	77.52	89.84	70.66
62.58	39.51	11.12	35.07
91.28	83.85	64.48	81.68
31.41	87.19	19.92	49.37
71.15	59.57	22.53	69.16
50.82	24.23	65.99	57.84
22.78	94.70	66.75	53.69
33.34	61.90	11.54	96.17
25.12	50.05	74.45	36.03
92.49	82.98	55.62	30.35
58.43	93.63	95.37	39.51
75.64	20.67	72.71	48.15

5. $22.78	6. $94.70	7. $66.75	8. $79.53
69.33	34.61	90.72	71.09
48.14	27.10	80.11	54.96
17.81	68.47	73.29	59.15
44.88	76.13	56.25	50.91
40.18	31.05	74.45	57.42
19.02	26.30	35.58	43.93
63.95	37.86	24.38	32.23
89.16	46.65	39.51	85.64
99.08	20.67	84.36	28.41
87.83	92.49	82.98	55.01
77.52	21.60	92.13	16.46
22.78	56.25	49.12	50.91
40.18	31.82	94.70	98.55
66.75	62.77	52.05	74.79
53.45	69.33	34.57	21.65
60.39	51.85	64.61	90.72
71.09	48.15	27.10	80.06

Exercise No. 35

Left-to-Right Subtraction

This exercise illustrates a principle: if a figure in the subtrahend is the same as the one above it in the minuend, the effect on the minuend will depend on whether or not a borrowing has been necessary with the next figure to the right.

In the first example we see that because 9 is greater than 4, the 5 in the minuend becomes a 4, and since 5 is greater than this the 7 in the minuend becomes a 6. We perform the subtraction thus: 3 from 6 leaves 3, 5 from 14 leaves 9, 9 from 14 leaves 5; answer, 395.

1. 754	2. 655	3. 251	4. 546	5. 592
359	358	159	247	294

6. 862	7. 444	8. 968	9. 773	10. 763
667	146	569	279	266

11. 832	12. 233	13. 983	14. 572	15. 656
536	139	488	278	357

16. 395	17. 856	18. 645	19. 721	20. 941
197	659	248	428	249

21. 527	22. 863	23. 985	24. 267	25. 843
329	569	389	168	448

Exercise No. 36

Trios that Add to 27 or Less

The groups of three here add to numbers between 21 and 27. Add by combining these groups. *Add from the top down.*

1. 36	2. 63	3. 47	4. 65	5. 47
98	79	87	78	97
99	89	98	98	99
69	86	74	87	75
99	89	78	87	78
99	89	79	99	89
56	33	67	54	49
89	99	77	89	89
89	99	97	99	99
73	67	84	77	75
79	97	88	87	78
99	97	99	88	78

6. 55	**7.** 68	**8.** 56	**9.** 68	**10.** 56
88	88	87	88	98
89	88	99	99	98
77	85	78	96	78
78	99	88	98	89
98	99	89	98	99
65	57	96	68	66
89	98	97	89	78
89	99	98	99	89
87	76	78	96	84
98	87	78	97	88
98	98	88	99	89

Exercise No. 37

Left-to-Right Subtraction

In these examples another consideration arises: the tens' figure in the minuend is 0; when 1 is borrowed to make possible the subtraction of the units, the tens in the minuend become 9 and the hundreds are also reduced by 1.

To illustrate with the first example: 3 from 6 leaves 3, 5 from 9 leaves 4, 7 from 14 leaves 7; answer, 347.

Subtract from left to right.

1. 704	**2.** 307	**3.** 806	**4.** 204	**5.** 404
357	118	457	126	297

6. 808	**7.** 706	**8.** 308	**9.** 302	**10.** 203
549	517	189	236	115

11. 800	**12.** 501	**13.** 300	**14.** 805	**15.** 601
585	323	122	796	374

16. 902	**17.** 500	**18.** 408	**19.** 700	**20.** 207
793	386	159	466	178

21. 807	**22.** 603	**23.** 200	**24.** 600	**25.** 300
509	319	162	224	171

Exercise No. 38
Adding Single Columns by Pairs
Take pairs of successive numbers at a time. *Add from the bottom up.*

1. $5759.37	**2.** $7856.21	**3.** $6525.49
2186.62	2477.50	5214.44
4491.67	5843.84	8788.76
3848.60	3993.36	1115.81
6874.79	4751.85	2740.32
1831.04	9213.53	4569.82
1080.33	3363.26	9528.30
6461.73	9994.90	7271.70
9823.34	9617.89	8983.55

4. $4142.97	**5.** $6675.01	**6.** $1916.46
4629.22	3508.07	2009.03
2089.83	5624.21	6538.82
9766.48	6039.10	8788.80
3367.72	7677.25	7531.01
9849.04	6393.03	8635.19
1623.26	6257.59	5096.58
4308.52	3646.51	1185.13
5354.34	9678.28	1714.55
4244.07	7170.27	4015.81
6874.79	3229.30	6422.37
6118.91	4569.73	9947.94

Exercise No. 39

Mental Subtraction

Use the method of making the subtrahend a round number. Subtract $1 from the minuend and add to this the difference between $1 and the given subtrahend.

Taking the first example: $1 from $5.18 leaves $4.18; $.83 from $1 leaves $.17; $4.18 + $.17 = $4.35.

1. $5.18 − $.83		11. $3.22 − $.93
2. $6.42 − $.83		12. $7.37 − $.61
3. $1.89 − $.95		13. $4.56 − $.97
4. $2.47 − $.99		14. $6.87 − $.91
5. $7.48 − $.56		15. $2.21 − $.65
6. $8.29 − $.66		16. $4.86 − $.97
7. $3.18 − $.87		17. $3.32 − $.64
8. $7.27 − $.43		18. $7.75 − $.83
9. $4.19 − $.49		19. $4.12 − $.63
10. $3.53 − $.77		20. $6.23 − $.26

Exercise No. 40

Adding Single Columns by Trios

Do the addition examples in Exercise No. 13 on page 11 by grouping three numbers at a time.

Taking the first example there presented, the following illustrates the method of adding: 13 (+12) 25, write 5 and carry 2; 2 (+17) 19, (+17) 36; answer, 365. Do not consciously repeat to yourself the individual amounts that you are adding, but only the successive total. *Add from the top down.*

Exercise No. 41

Adding Single Columns by Pairs

1. $7489.99	2. $8356.24	3. $2165.38
2897.66	4860.39	1034.96
7828.17	8084.05	8788.86
3519.16	2303.32	2922.64
2237.61	1891.45	4142.44
7170.27	4015.94	9062.57
5950.95	5843.08	9849.04
1209.63	9326.73	4768.79
8152.92	3646.51	1185.13
5354.14	5520.33	6772.76
7725.75	3104.60	1348.37
6101.98	4953.91	6039.62
5429.30	6772.76	1780.84
4414.57	5910.18	9134.96
7812.07	7170.06	8788.86
5056.24	9564.22	7755.63
2593.26	2075.27	4033.03
4569.35	9236.74	8932.58

4. $8799.55	5. $1319.16	6. $8348.84
4437.14	5781.63	2538.82
9793.08	5266.88	2861.41
4223.59	3926.73	9809.50
3218.94	9156.24	5834.43
9564.65	2227.49	5340.33
6296.78	1207.54	5446.31
4569.35	7729.30	5115.71
7006.68	6772.11	8521.65
7976.92	9036.17	8074.89
3612.97	8909.50	2124.56
8765.77	2930.51	1507.23
5960.54	9964.75	2279.76
5546.31	7188.86	2858.34
4347.04	4147.61	8085.37
9570.06	1457.10	4884.44
6935.05	3218.94	8168.39
6774.27	4913.26	7273.93

Exercise No. 42

Mental Subtraction

Perform the subtractions in Exercise No. 39 by using the method of making a round number of the minuend. That is, reduce the minuend to the next lower number of even dollars. Subtract the subtrahend from this and then add the excess of cents in the minuend.

Taking the first example ($5.18 − $.83): $.83 from $5 leaves $4.17; $4.17 + 18 = $4.35.

Exercise No. 43

Mental Subtraction

Perform the following subtractions mentally. Raise the subtrahend to the next larger number of even dollars.

1. $2.79 − $1.86
2. $3.17 − $1.97
3. $9.50 − $6.69
4. $2.56 − $1.91
5. $4.77 − $2.81
6. $9.78 − $3.94
7. $7.44 − $4.49
8. $4.37 − $2.72
9. $5.22 − $2.98
10. $6.04 − $5.33

11. $5.53 − $3.64
12. $2.62 − $1.89
13. $3.05 − $1.82
14. $8.28 − $6.65
15. $8.10 − $6.39
16. $5.15 − $2.67
17. $4.47 − $2.61
18. $7.93 − $5.99
19. $5.40 − $2.95
20. $3.23 − $1.60

Exercise No. 44

Mental Subtraction

Do the examples in Exercise No. 43 by lowering the minuend to the next smaller number of even dollars.

MULTIPLICATION IN GENERAL

Multiplication is the heart's core of the art of calculation. In itself it constitutes an art about which a large volume might be written.

The multiplication exercises in this book have three main objects in view—first, to enable the student to use all numbers up to 25 as direct multipliers in written work; second, to teach him to multiply mentally any number up to 1000 by any other number up to 1000; third, to drill him in various short-cut methods that apply to particular cases.

The use of numbers up to 25 as direct multipliers may be illustrated by this example:

A	B
7648	7648
1923	1923
22944	175904
15296	145312
68832	14707104
7648	
14707104	

In Method A, which is here shown for comparison, the usual procedure is followed. In Method B the calculation is performed thus: $8 \times 23 = 184$, write 4 and carry 18; $4 \times 23 = 92$, $92 + 18 = 110$, write 0 and carry 11; $6 \times 23 = 138$, $138 + 11 = 149$, write 9 and carry 14; $7 \times 23 = 161$, $161 + 14 = 175$. Multiplication by 19 is done in the same way, and the partial products added.

To multiply in the manner described it is of course necessary to acquire a knowledge of the multiplication table up to 25 × 25. Instruction in this direction is given by very easy steps. There are several types of exercises leading to the same end.

Exercises in mental multiplication are similarly graded. You start by multiplying two figures by one, then two by two, then three by one, three by two, and finally three by three.

The subject of short cuts is highly specialized and need not detain us for the present.

Exercise No. 45
Mental Multiplication

Multiply by 2 the numbers in Table I on page 7. Proceed from left to right. A few examples of the method calculating will suffice.

32 × 2: 30 × 2 = 60, 2 × 2 = 4, 60 + 4 = 64
45 × 2: 40 × 2 = 80, 5 × 2 = 10, 80 + 10 = 90
49 × 2: 40 × 2 = 80, 9 × 2 = 18, 80 + 18 = 98
99 × 2: 90 × 2 = 180, 9 × 2 = 18, 180 + 18 = 198

Exercise No. 46
Mental Multiplication

Multiply mentally by 3 the numbers in Table I on page 7.

Exercise No. 47
Mental Multiplication

Multiply mentally by 4 the numbers in Table I on page 7.

Exercise No. 48
Adding Single Columns by Pairs
Take pairs of successive numbers at a time. *Add from the bottom up.*

1. $227976.55
   491368.39
   476170.02
   804501.33
   920950.63
   512573.15

2. $364631.71
   291241.97
   620314.57
   378990.83
   267278.30
   586721.69

3. $693505.74
   822427.23
   186620.98
   871060.54
   118577.94
   996475.17

4. $430413.93
   525632.59
   198886.28
   651653.40
   964295.81
   480444.80

5. $605465.38
   599320.95
   810064.74
   112279.76
   431275.17
   890890.55

6. $694235.68
   483929.91
   841653.40
   344518.66
   624133.37
   364698.97

Exercise No. 49

Mental Subtraction

Raise the subtrahend to the next larger number of even dollars.

1. $19.03 − $.50	9. $61.70 − $.94
2. $26.52 − $.86	10. $72.04 − $.85
3. $24.27 − $.32	11. $67.30 − $.73
4. $15.58 − $.80	12. $60.54 − $.69
5. $42.35 − $.59	13. $94.20 − $.48
6. $39.29 − $.91	14. $81.64 − $.74
7. $16.53 − $.79	15. $76.34 − $.66
8. $43.12 − $.17	16. $62.41 − $.89

Exercise No. 50

Mental Multiplication

Multiply mentally by 5 the numbers in Table I on page 7.

Exercise No. 51

Mental Subtraction

Do the examples in Exercise No. 49 by reducing the minuend to the next smaller number of even dollars.

Exercise No. 52

Mental Multiplication

Multiply mentally by 6 the numbers in Table I on page 7.

Exercise No. 53

Mental Multiplication

Multiply mentally by 7 the numbers in Table I on page 7.

Exercise No. 54
Adding Single Columns by Pairs
Take pairs of successive numbers at a time. *Add from the top down.*

1. $806054.65
   681097.85
   451866.93
   431248.39
   298291.24
   322157.61
   700177.25
   714913.58
   746789.23
   569055.36
   534011.98
   281472.87

2. $386942.35
   933492.59
   209507.09
   751706.02
   882750.78
   305181.62
   733115.33
   379499.64
   663265.52
   444684.16
   227976.86
   377730.32

3. $243130.39
   158010.21
   519794.95
   893672.07
   870485.02
   834913.40
   287919.76
   697537.73
   225942.35
   435756.84
   996168.05
   164864.14

4. $559663.93
   882067.60
   265254.65
   332750.44
   380353.71
   462925.62
   583492.78
   411711.98
   230882.09
   911270.45
   180190.66
   744732.86

Exercise No. 55
Mental Subtraction

Raise the subtrahend to the next larger number of even dollars.

1. $24.31 − $4.55
2. $26.36 − $7.50
3. $49.13 − $4.62
4. $34.37 − $7.98
5. $43.12 − $1.70
6. $14.06 − $7.86
7. $15.10 − $2.88
8. $26.52 − $6.89

9. $96.15 − $8.88
10. $87.04 − $2.53
11. $79.19 − $7.58
12. $59.42 − $3.82
13. $99.05 − $1.90
14. $77.24 − $3.55
15. $67.60 − $5.97
16. $72.07 − $3.87

Exercise No. 56
Mental Multiplication

Multiply mentally by 8 the numbers in Table I on page 7.

Exercise No. 57
Adding Single Columns by Trios

Do the examples in Exercise No. 15 on page 12 by taking three successive numbers at a time. *Add from the top down.*

Exercise No. 58
Mental Subtraction

Do the examples in Exercise No. 55 by lowering the minuend to the next smaller number of even dollars.

Exercise No. 59
Addition of Partial Products

The type of exercise here presented has a bearing on mental multiplication. Thus the first example represents, in inverted position, the partial products we get when we multiply 15 by 53.

$$\begin{array}{r} 15 \\ 53 \\ \underline{45} \\ 750 \\ \underline{} \\ 795 \end{array}$$

When partial products of this kind occur in mental multiplication you are of necessity compelled *to retain them in your mind*. Hence to develop your ability to do this kind of memory work, you are asked to read each example once and then write it three times on paper before you perform the mental addition.

Complete the mental addition before writing the answer. Work from left to right. Thus in doing the first example you would say to yourself: 750, 790, 795. In doing the second you would say: 620, 680, 682.

1. 750	**2.** 620	**3.** 470	**4.** 740	**5.** 520
45	62	94	74	78
6. 880	**7.** 720	**8.** 880	**9.** 960	**10.** 840
44	90	66	72	72
11. 850	**12.** 540	**13.** 570	**14.** 220	**15.** 910
51	81	95	88	52
16. 680	**17.** 980	**18.** 280	**19.** 640	**20.** 690
34	28	84	96	92
21. 760	**22.** 810	**23.** 750	**24.** 910	**25.** 580
95	54	15	78	87

Exercise No. 60
Mental Multiplication
Multiply mentally by 9 the numbers in Table I on page 7.

Exercise No. 61

Mental Multiplication

Multiply mentally by 11 the numbers in Table I.

Exercise No. 62

Adding Single Columns by Pairs

Add from the bottom up.

1. $698504.99	2. $457012.91
845643.09	820823.58
761979.28	622529.46
401349.83	715303.47
740614.80	159363.96
553930.31	380272.36
896554.52	268195.94
975160.67	789234.17
417337.75	773286.20
882110.35	425922.98
116448.16	669001.18
477406.66	502733.07
801415.93	906396.55
340939.01	301831.05
380272.36	820889.23
656958.68	548620.61
882152.17	874185.10
401304.99	761944.26

3.	$662533.75	4.	$473105.74
	380277.80		141593.51
	847236.82		111290.63
	735356.57		897350.27
	236569.58		379128.68
	862061.88		966221.52
	178735.81		644107.29
	464385.34		104004.99
	425919.44		266722.95
	789249.94		987983.35
	395497.48		183216.70
	194426.67		295788.92
	129066.25		336353.75
	464347.56		578389.73
	316085.34		740638.09
	499498.27		236540.02
	776980.14		159383.58
	518437.35		729128.36

Exercise No. 63

Mental Subtraction

Raise the subtrahend to the next larger number of even dollars.

1. $83.37 − $35.72
2. $68.20 − $61.99
3. $97.48 − $17.87
4. $64.41 − $29.67

5. $25.33 − $10.65
6. $79.58 − $51.84
7. $48.54 − $20.61
8. $52.17 − $30.32

9. $91.28 − $36.82
10. $76.42 − $62.59
11. $55.30 − $18.81
12. $95.12 − $90.66

13. $65.40 − $14.93
14. $37.35 − $28.82
15. $49.01 − $21.85
16. $81.03 − $41.16

Exercise No. 64

Continuous Addition Drill

Count by 3's to 75.
Count by 4's to 100.
Count by 6's to 150.
Count by 7's to 175.
Count by 8's to 200.
Count by 9's to 225.
Count by 11's to 275.
Count by 12's to 300.

Repeat this exercise three times.

Exercise No. 65

Mental Subtraction

Do the examples in Exercise No. 63 by lowering the minuend to the next smaller number of even dollars.

Exercise No. 66

Mental Addition

Read each of these examples once, write it three times and then add it mentally from left to right.

Be careful to think of the upper number in each case as something in the thousands and not as so many hundreds. Thus in the first example the upper number should be called one thousand seven hundred forty, *not* seventeen hundred forty. It is easier to think of comparatively small numbers as hundreds rather than as thousands plus hundreds, but this method of naming leads to trouble when dealing with larger numbers, and it is best to follow one uniform system.

1. 1740	2. 1650	3. 1080	4. 1280
87	55	90	96

5. 2430	6. 2560	7. 3690	8. 1120
81	64	82	80

9. 1450	10. 1140	11. 1320	12. 1350
87	95	88	78

13. 1340	14. 1320	15. 1920	16. 2340
67	88	96	78

17. 3680	18. 1080	19. 1950	20. 2520
92	84	65	72

Exercise No. 67
Mental Subtraction

Raise the subtrahend to the next larger number of even dollars.

1. $855.30 − $8.32
2. $844.16 − $7.29
3. $671.46 − $4.47
4. $834.06 − $4.09
5. $642.02 − $7.80
6. $836.11 − $8.68
7. $862.21 − $4.45
8. $532.13 − $4.41

9. $426.22 − $7.78
10. $912.25 − $5.33
11. $453.31 − $5.60
12. $594.10 − $7.23
13. $415.37 − $7.91
14. $520.39 − $9.76
15. $542.17 − $8.55
16. $673.29 − $9.44

Exercise No. 68
Adding Single Columns by Trios

Do the examples in Exercise No. 17 on page 15 by grouping three successive numbers at a time. *Add from the top down.*

Exercise No. 69
Mental Subtraction

Do the examples in Exercise No. 67 by reducing the minuend to the next smaller number of even dollars.

Table II
Numbers for Multiplication Table Drill

A	B	C	D	E	F	G	H	J	K	L	M
2	2	2	2	2	2	2	2	2	2	2	2
4	5	6	7	8	9	10	11	8	9	10	11
6	8	10	12	14	16	18	20	14	16	18	20
8	11	14	17	3	3	3	3	20	23	3	3
10	14	3	3	9	10	11	12	13	3	11	12
12	3	7	8	15	17	19	21	9	10	19	21
14	6	11	13	4	4	4	4	15	17	4	4
3	9	15	4	10	11	12	13	21	4	12	13
5	12	4	9	16	18	20	5	4	11	20	22
7	15	8	14	5	5	5	14	10	18	5	5
9	4	12	5	11	12	13	6	16	5	13	14
11	7	16	10	17	19	6	15	22	12	21	23
13	10	5	15	6	6	14	7	5	19	6	6
	13	9	6	12	13	7	16	11	6	14	15
		13	11	18	7	15	8	17	13	22	24
			16	7	14	8	17	6	20	7	7
				13	8	16	9	12	7	15	16
					15	9	18	18	14	23	25
						17	10	7	21	8	8
							19	13	8	16	17
								19	15	24	9
									22	9	18
										17	10
											19

Exercise No. 70

Multiplication Table Drill

Use Table II on this page. Multiply the numbers in Column A successively by 2, 3, 4, 5, 6, 7, 8, 9, 10, 11, and 12. Repeat this exercise three times.

Exercise No. 71

Mental Subtraction

Raise the subtrahend to the next larger number of even dollars, and raise this amount in turn to an even $100. Thus, taking the first example: $100 from $365.42 leaves $265.42; $265.42 + $11 (difference between $100 and $89) equals $276.42; $276.42 + $.27 = $276.69.

1. $365.42 − $88.73	9. $459.48 − $87.55
2. $950.49 − $94.98	10. $553.18 − $81.64
3. $723.67 − $40.77	11. $416.07 − $29.19
4. $614.15 − $93.79	12. $426.22 − $95.78
5. $858.51 − $84.72	13. $912.25 − $33.63
6. $928.36 − $36.82	14. $753.46 − $56.57
7. $413.54 − $86.61	15. $831.05 − $60.85
8. $342.21 − $96.62	16. $743.16 − $68.29

Exercise No. 72

Adding Single Columns by Trios

Do the examples in Exercise No. 22 on page 20 by grouping three successive numbers at a time. *Add from the bottom up.*

Table III

Numbers to Be Multiplied

1. 111315	6. 171922	11. 222572
2. 111417	7. 182123	12. 541418
3. 121416	8. 897254	13. 192389
4. 121518	9. 248963	14. 151924
5. 541316	10. 258163	15. 212481

Exercise No. 73
Written Multiplication
Multiply the numbers in Table III by 6789.

Exercise No. 74
Mental Addition
Read each of the following examples once, write it three times and then add it mentally from left to right.

Think of the upper number in each case as being in the thousands and not the hundreds.

The first example would be added: 1280, 1480, 1536. In other words, take the first number as a whole, and then add to it successively the hundreds, tens and units of the second number.

1. 1280	2. 4410	3. 1960	4. 1380
256	196	686	115

5. 4620	6. 3060	7. 6510	8. 4150
693	170	837	664

9. 4080	10. 1110	11. 6480	12. 1450
204	185	144	174

13. 1640	14. 3350	15. 5150	16. 3510
246	268	344	351

17. 3040	18. 8080	19. 1240	20. 2250
304	528	372	405

Exercise No. 75

Mental Subtraction

Do the examples in Exercise No. 71 on page 49 by lowering the minuend. Reduce it to the next smaller number of even dollars. Taking the first example: $300 − $88.73 leaves $211.27; $211.27 + $65 = $276.27; $276.27 + $.42 = $276.69.

Exercise No. 76

Adding Single Columns by Trios

Do the examples in Exercise No. 26 on page 23 by grouping three successive numbers at a time. *Add from the top down.*

Exercise No. 77

Mental Multiplication

Multiply mentally by 12 the numbers in Table I on page 7.

Exercise No. 78

Adding Single Columns by Trios

Do the examples in Exercise No. 34 on page 28 by grouping three successive numbers at a time.

Exercise No. 79

Mental Subtraction

Raise the subtrahend to the next larger number of even hundreds of dollars.

1. $950.49 − $498.65	5. $769.14 − $580.93
2. $646.43 − $456.57	6. $831.05 − $685.34
3. $520.39 − $176.42	7. $821.45 − $529.48
4. $821.13 − $468.54	8. $862.39 − $197.76

9. $318.32 − $181.64   13. $416.07 − $219.44
10. $636.09 − $549.95   14. $640.02 − $493.79
11. $714.10 − $273.65   15. $746.14 − $159.93
12. $821.45 − $599.97   16. $752.30 − $183.81

Exercise No. 80

Mental Addition

Read each of the following examples once, write it three times and then add it mentally from left to right. The first example would be added: 16530, 17030, 17081.

1. 16530	2. 12930	3. 24920
551	431	623

4. 22080	5. 37150	6. 33650
552	743	673

7. 51780	8. 44460	9. 67340
863	741	962

10. 61810	11. 19360	12. 12160
883	242	152

13. 76960	14. 32670	15. 25380
962	363	282

16. 12690	17. 15320	18. 19620
141	766	654

19. 21720	20. 46650	21. 44160
543	933	736

Exercise No. 81

Written Multiplication

Multiply by 1112 each of the numbers in Table III on page 49. Wherever there occurs in the multiplicand a pair of figures that may be considered as 11 or 12, make one multiplication of this instead of two, and accordingly write down two figures in the partial product. Taking the first example:

$$
\begin{array}{r}
111315 \\
1112 \\
\hline
1335780 \\
1224465 \\
\hline
123782280
\end{array}
$$

111315 is successively multiplied (from right to left) by 12 and 11 thus: $5 \times 12 = 60$, write 0 and carry 6; $1 \times 12 = 12$, $12 + 6 = 18$, write 8 and carry 1; $3 \times 12 = 36$, $36 + 1 = 37$, write 7 and carry 3; $11 \times 12 = 132$, $132 + 3 = 135$, write 35 and carry 1; $1 \times 12 = 12$, $12 + 1 = 13$, write 13. Multiplication by 11 is carried out in the same way.

In doing these examples be watchful about placing the second partial product *two* places to the left of the first.

Exercise No. 82

Adding Single Columns by Trios

Do the examples in Exercise No. 38 on page 32 by grouping three successive numbers at a time. *Add from the bottom up.*

Exercise No. 83

Mental Subtraction

Do the examples in Exercise No. 79 on page 51 by lowering the minuend to the next smaller number of even hundreds of dollars.

Exercise No. 84

Mental Addition

Read each of the following examples once, write it three times and then add it mentally from left to right.

Add in turn the thousands, hundreds, tens and units to the upper number. In doing the first example you should say to yourself something like the following: 18360 + 1224, 19360; 19360 + 224, 19560; 19560 + 24, 19584.

1. 18360 1224	**2.** 21630 2163	**3.** 24960 3328
4. 18820 5646	**5.** 16260 1084	**6.** 19530 1953
7. 21360 2848	**8.** 16420 4926	**9.** 18640 6524
10. 10290 2401	**11.** 13530 3608	**12.** 16860 5058
13. 29240 1462	**14.** 33680 2526	**15.** 28590 4765
16. 13230 3969	**17.** 26520 1326	**18.** 28840 2163
19. 24960 4160	**20.** 28290 5658	**21.** 14120 2118

Exercise No. 85

Continuous Addition Drill

Count by 4's to 100.

Count by 6's to 150.

Count by 7's to 175.

Count by 8's to 200.

Count by 9's to 225.

Count by 11's to 275.

Count by 12's to 300.

Count by 13's to 325.

Repeat this exercise three times.

Exercise No. 86

Adding Single Columns by Trios

Do the examples in Exercise No. 41 on page 34 by grouping three successive numbers at a time. *Add from the top down.*

Exercise No. 87

Factoring

When numbers are multiplied together, they are considered *factors* of the resulting *product*. Thus 2 and 3 are factors of 6, and 3 and 5 are factors of 15.

Factoring a number is the process of resolving the number into the factors that will produce the number when multiplied together. Thus 36 may be factored as 2×18, or as 3×12, or as 4×9, or as 6×6.*

Any number that can be resolved into factors is called a *composite* number.

A *prime* number is one that has no factors besides itself and 1. Thus, 1, 2, 3, 5, 7, 11, 13, etc. are prime numbers.

---

* If it were required to give the *prime* factors of 36, these would be $2 \times 2 \times 3 \times 3$, but factoring into prime numbers has nothing to do with the purposes of this book.

On the pages starting with 146 will be found a table which analyzes all prime and composite numbers up to 625. You will be taught gradually to familiarize yourself with this entire table. The purpose of this is to help you to recognize quickly the character of these numbers—to enable you to multiply rapidly the factors that produce any of them, or to separate any of them into such factors.

Of special importance in this table are the numbers printed in italic type, since these can be produced by two factors each of which is 25 or less.

It is quite commonly appreciated that very small numbers have a definite individuality which grows out of the many associations built up around them in our minds. The individual character of higher numbers becomes similarly apparent and unforgettable when we single them out for particular attention.

For the first exercise in factoring read the first two columns of the table on page 146, and then write these from memory (or calculation) in the same form.

In studying the table note that each composite number is factored by first taking the smaller factors in the order of their size, and that the combinations are not repeated. Thus the separate ways of factoring 48 are given as 2×24, 3×16, 4×12 and 6×8. These combinations are not repeated as 8×6, 12×4, 16×3, and 24×2.

Exercise No. 88

Multiplication Table Drill

Use Table II on page 48.

Multiply the numbers in Column A successively by 3, 4, 6, 7, 8, 9, 11, 12 and 13.

Repeat this exercise three times.

This exercise takes us the first step beyond the customary limits of the multiplication table, which ordinarily goes no farther than 12 × 12. Succeeding examples will enable you to memorize the products of all pairs of numbers up to 25 × 25.

No multiplication table, as such, is presented in this book, because learning the products of higher factors by sheer power of memory is extremely difficult. On the other hand, when you are put over and over again to the necessity of figuring out these higher combinations for yourself, they soon come to stick firmly in the mind.

Exercise No. 89

Mental Addition

Read each of the following examples once, write it three times, and then add it mentally from left to right. The first example would be added: 165300, 170300, 170810.

1. 165300	**2.** 129300	**3.** 249200
5510	4310	6230
4. 220800	**5.** 371500	**6.** 336500
5520	7430	6730
7. 517800	**8.** 444600	**9.** 673400
8630	7410	9620
10. 618100	**11.** 193600	**12.** 121600
8830	2420	1520
13. 769600	**14.** 326700	**15.** 253800
9620	3630	2820

| **16.** 126900 | **17.** 153200 | **18.** 196200 |
| 1410 | 7660 | 6540 |

| **19.** 217200 | **20.** 456500 | **21.** 441600 |
| 5430 | 9330 | 7360 |

Exercise No. 90
Mental Multiplication

Multiply mentally by 13 the numbers in Table I on page 7.

In working with numbers from 80 upward, immediately name 1000 as the first part of the product. Thus 83 × 13 is 1040, (+39) 1079; 97 × 13 is 1170, 1261.

Exercise No. 91
Adding Single Columns by Trios

Do the examples in Exercise No. 48 on page 39 by grouping three successive numbers at a time. *Add from the bottom up.*

Exercise No. 92
Factoring

Read the table on page 146 from 31 to 72 inclusive, and then write it in the same form.

Exercise No. 93
Mental Addition

Read each of the following examples once, write it three times and then add it mentally from left to right.

Add in turn the tens of thousands, thousands, hundreds and tens to the upper number. The first example would be added: 183600, 193600, 195600, 195840.

| 1. 183600 | 2. 216300 | 3. 249600 |
| 12240 | 21630 | 33280 |

| 4. 188200 | 5. 162600 | 6. 195300 |
| 56460 | 10840 | 19530 |

| 7. 213600 | 8. 164200 | 9. 186400 |
| 28480 | 49260 | 65240 |

| 10. 102900 | 11. 135300 | 12. 168600 |
| 24010 | 36080 | 50580 |

| 13. 292400 | 14. 336800 | 15. 285900 |
| 14620 | 25260 | 47650 |

| 16. 132300 | 17. 265200 | 18. 288400 |
| 39690 | 13260 | 21630 |

| 19. 249600 | 20. 282900 | 21. 141200 |
| 41600 | 56580 | 21180 |

Exercise No. 94

Written Multiplication

Multiply by 1213 each of the numbers in Table III on page 49. Wherever there occurs in the multiplicand a pair of figures that may be considered as 11, 12 or 13, make one multiplication of this instead of two, and write two figures in the partial product. Thus, taking the first example, we successively multiply 15, 13 and 11 by 13 and again by 12. The partial products are accordingly written in two lines instead of the customary four.

Exercise No. 95

Adding Single Columns by Trios

Do the examples in Exercise No. 54 on page 41 by grouping three successive numbers at a time. *Add from the top down.*

Exercise No. 96

Factoring

Factor the numbers from 54 to 92 inclusive in the form shown in the table on page 146.

Exercise No. 97

Mental Addition

Read each of the following examples once, write it three times and then add it mentally from left to right.

Add the whole of the second number to the first before considering the third. Repeat to yourself several times the sum of the first and second if you find this necessary.

The third example would be added: 36300, 39300, 39930; (repeat 39930, 39930); 39930, 40030, 40051.

1. 10100	2. 22200	3. 36300
1010	2220	3630
101	222	121

4. 52400	5. 70500	6. 90600
5240	7050	1510
262	141	302

7. 19100	8. 20200	9. 33300
9950	1010	2220
382	101	222

10. 48400	11. 65500	12. 84600
3630	5240	7050
121	262	141

13. 18100	14. 38200	15. 20200
7240	9050	4040
181	905	202

16. 42400	17. 66600	18. 40400
6360	8880	4040
424	666	404

19. 33600	20. 88800	21. 30300
3360	8880	9090
336	222	303

Exercise No. 98
Continuous Addition Drill
Count by 6's to 150.
Count by 7's to 175.
Count by 8's to 200.
Count by 9's to 225.
Count by 11's to 275.
Count by 12's to 300.
Count by 13's to 325.
Count by 14's to 350.

Repeat this exercise three times.

Exercise No. 99

Adding Single Columns by Trios

Do the examples in Exercise No. 62 on page 44 by grouping three successive numbers at a time. *Add from the bottom up.*

Exercise No. 100

Factoring

Factor the numbers from 73 to 111 inclusive in the form shown in the table on page 146.

Exercise No. 101

Mental Addition

Read each of the following examples once, write it three times and then add it mentally from left to right.

The first example would be added: 26200, 33200, 34000, 34060; 34060, 36060, 36156.

1. 26200	**2.** 48400	**3.** 69900
7860	9680	9320
2096	1210	1398
4. 12100	**5.** 26400	**6.** 42900
9680	9240	8580
1089	1056	1144
7. 61600	**8.** 82500	**9.** 88000
9240	9900	8800
1078	1155	1056
10. 93500	**11.** 98000	**12.** 73200
9350	9800	9760
1122	1188	1098
13. 93100	**14.** 97600	**15.** 71000
9310	9760	7100
1064	1220	1065
16. 46600	**17.** 57700	**18.** 68800
9320	5770	6880
1398	2308	2064

19. 79900	**20.** 24600	**21.** 70200
7990	9840	9320
3196	1107	1170

Exercise No. 102

Multiplication Table Drill

Use Table II on page 48.

Multiply the numbers in Column A successively by 4, 6, 7, 8, 9, 11, 12, 13 and 14.

Repeat this exercise three times.

Exercise No. 103

Two-Column Addition

You are now ready to start adding two columns at a time. Take Exercise No. 13 on page 11. *Add from the top down.*

Two-column addition is simply an application of the left-to-right methods which you have already learned. To illustrate with the first example:

$$43$$
$$62$$
$$78$$
$$81$$
$$14$$
$$\underline{87}$$

This would be added: 43, 103, 105, 175, 183, 263, 264, 274, 278, 358, 365. These are the actual steps, but with practice you will read this as 105, 183, 264, 278, 365.

Exercise No. 104

Factoring

Factor the numbers from 93 to 129 inclusive in the form shown in the table on pages 146 and 147.

Exercise No. 105

Mental Addition

Read each of the following examples once, write it three times, and then add it mentally from left to right.

1. 112700 3220 161	**2.** 136800 5130 342	**3.** 162900 2400 181
4. 105700 1510 302	**5.** 128800 3220 161	**6.** 153900 5130 342
7. 151200 5040 756	**8.** 183400 7860 262	**9.** 176400 5040 252
10. 209600 7860 524	**11.** 104800 5240 524	**12.** 103200 6880 860
13. 114100 6520 978	**14.** 112800 7050 423	**15.** 126000 7560 756
16. 111000 9250 740	**17.** 104400 8700 870	**18.** 135900 9060 302
19. 112800 9870 141	**20.** 130500 8700 435	**21.** 136800 6800 684

Exercise No. 106

Mental Multiplication

Multiply mentally by 14 the numbers in Table I on page 7.

Exercise No. 107

Two-Column Addition

Do the examples in Exercise No. 17 on page 15 by adding two columns at a time. *Add from the bottom up.*

Exercise No. 108

Factoring

Factor the numbers from 112 to 145 inclusive in the form shown in the table on pages 146 and 147.

Exercise No. 109

Mental Addition

Read each of the following examples once, write it three times, and then add it mentally from left to right.

1. 121000	**2.** 217600	**3.** 253800
14520	10880	14100
484	544	846
4. 116000	**5.** 145200	**6.** 224800
11600	14520	10880
464	726	816
7. 171500	**8.** 211800	**9.** 344700
24010	10590	22980
343	706	383
10. 129200	**11.** 166500	**12.** 290400
16150	19980	14520
323	666	363

13. 335700	**14.** 272400	**15.** 324800
18650	18160	23200
746	454	928

16. 124200	**17.** 317800	**18.** 371200
20700	18160	23200
828	454	924

19. 395500	**20.** 210000	**21.** 540800
34200	36750	33800
565	525	676

Exercise No. 110
Written Multiplication

Multiply by 1314 the numbers in Table III on page 49.

Exercise No. 111
Two-Column Addition

Do the examples in Exercise No. 26 on page 23 by adding two columns at a time. *Add from the top down.*

Exercise No. 112
Factoring

Factor the numbers from 130 to 162 inclusive in the form shown in the table on page 147.

Exercise No. 113
Mental Addition

Read each of the following examples once, write it three times, and then add it mentally from left to right.

1. 123200	**2.** 187800	**3.** 254400
39800	37560	44520
1232	1878	2544

4. 323000	5. 393600	6. 466200
51680	59040	26640
3230	3936	4662

7. 616200	8. 121200	9. 184800
41160	48480	55440
1392	2424	3080

10. 250400	11. 318000	12. 387600
25040	31800	38760
3956	4452	1292

13. 439200	14. 532800	15. 608400
43920	53280	60840
1312	1998	2704

16. 139200	17. 143400	18. 218700
34800	28680	36350
1392	1434	2187

19. 294800	20. 373500	21. 454200
44220	52290	60560
2948	3735	4542

Exercise No. 114

Continuous Addition Drill

Count by 7's to 175.
Count by 8's to 200.
Count by 9's to 225.
Count by 11's to 275.
Count by 12's to 300.
Count by 13's to 325.

Count by 14's to 350.
Count by 15's to 375.
Repeat this exercise three times.

Exercise No. 115

Two-Column Addition

Do the examples in Exercise No. 34 on page 28 by adding two columns at a time. *Add from the bottom up.*

Exercise No. 116

Multiplication Table Drill

Use Table II on page 48.

Multiply the numbers in Column B successively by 6, 7, 8, 9, 11, 12, 13, 14 and 15.

Repeat this exercise three times.

Exercise No. 117

Factoring

Factor the numbers from 146 to 179 inclusive in the form shown in the table on page 147.

Exercise No. 118

Two-Column Addition

Do the examples in Exercise No. 38 on page 32 by adding two columns at a time. *Add from the top down.*

It slows up addition by two columns to keep repeating the number of hundreds as you go along. A good plan is to keep tally of the number of hundreds with a pencil. In all addition of long columns write numbers to be carried either at the head of the next column or beneath the figures in the total as you set them down. When looking for errors in addition, add in the opposite direction from that in which the addition was originally performed.

Exercise No. 119

Mental Multiplication

Multiply mentally by 15 the numbers in Table I on page 7.

Exercise No. 120

Two-Column Addition

Do the examples in Exercise No. 41 on page 34 by adding two columns at a time. *Add from the bottom up.*

Exercise No. 121

Factoring

Factor the numbers from 163 to 194 inclusive in the form shown in the table on page 147.

Exercise No. 122

Two-Column Addition

Do the examples in Exercise No. 48 on page 39 by adding two columns at a time. *Add from the top down.*

Exercise No. 123

Written Multiplication

Multiply by 1415 the numbers in Table III on page 49.

Exercise No. 124

Two-Column Addition

Do the examples in Exercise No. 54 on page 41 by adding two columns at a time. *Add from the bottom up.*

Exercise No. 125
Factoring

Factor the numbers from 180 to 209 inclusive in the form shown in the table on page 147.

Exercise No. 126
Two-Column Addition

Do the examples in Exercise No. 62 on page 44 by adding two columns at a time. *Add from the top down.*

Exercise No. 127
Continuous Addition Drill
Count by 8's to 200.
Count by 9's to 225.
Count by 11's to 275.
Count by 12's to 300.
Count by 13's to 325.
Count by 14's to 350.
Count by 15's to 375.
Count by 16's to 400.
Repeat this exercise three times.

Exercise No. 128
Three-Column Addition
With the practice you have had in two-column addition you should now be able to add three columns at a time. Try this with the examples in Exercise No. 38 on page 32. No additional exercises in three-column addition are given, but you can of course practice it on your own account if you so desire.

Exercise No. 129
Multiplication Table Drill
Use Table II on page 48.
Multiply the numbers in Column C successively by 7, 8,
9, 11, 12, 13, 14, 15 and 16.
Repeat this exercise three times.

Exercise No. 130
Factoring
Factor the numbers from 195 to 224 inclusive in the
form shown in the table on pages 147 and 148.

Exercise No. 131
Mental Multiplication
Multiply mentally by 16 the numbers in Table I on
page 7 .

Exercise No. 132
Written Multiplication
Multiply by 1516 the numbers in Table III on page 49.

Exercise No. 133
Factoring
Factor the numbers from 210 to 239 inclusive in the
form shown in the table on pages 147 and 148.

DIVISION IN GENERAL

Division is multiplication in reverse. As you improve in multiplication you automatically develop your skill at division. For this reason it has been considered unnecessary to include any exercises in long division.

Exercises, however, are given in mental division, in order to round out your general calculating ability. These exercises are of the following types:

First you use the numbers from 2 to 25 as direct divisors, securing quotients from 1 to 99. Then you divide by the numbers from 2 to 9, finding answers of three places. Again, you divide by three-place numbers to arrive at quotients of one figure plus a remainder; the remainder is included so that the answer cannot be guessed but must be calculated accurately. Finally, you divide by numbers of two places and get results of two places. As division is somewhat more complicated, the exercises in division are not carried so far as those in multiplication.

Exercise No. 134
Mental Division

Divide mentally by 2 the answers to Exercise No. 45 as given on pages 161 and 162. Compare your answers with Table I on page 7.

Exercise No. 135
Continuous Addition Drill

Count by 9's to 225.
Count by 11's to 275.

Count by 12's to 300.
Count by 13's to 325.
Count by 14's to 350.
Count by 15's to 375.
Count by 16's to 400.
Count by 17's to 425.

Repeat this exercise three times.

Exercise No. 136
Mental Division

Divide mentally by 3 the answers to Exercise No. 46 as given on page 162. Compare your answers with Table I on page 7 .

Exercise No. 137
Multiplication Table Drill

Use Table II on page 48.

Multiply mentally the numbers in Column D by 8, 9, 11, 12, 13, 14, 15, 16 and 17.

Repeat this exercise three times.

Exercise No. 138
Factoring

Factor the numbers from 225 to 254 inclusive in the form shown in the table on page 148.

Exercise No. 139
Mental Division

Divide mentally by 4 the answers to Exercise No. 47 as given on page 162. Compare your answers with Table I on page 7 .

Exercise No. 140
Mental Multiplication

Multiply mentally by 17 the numbers in Table I on page 7 .

Exercise No. 141

Written Multiplication

Multiply by 1617 the numbers in Table III on page 49. Make a single multiplication of pairs of figures in the multiplicand up to 17.

Exercise No. 142

Factoring

Factor the numbers from 240 to 269 inclusive in the form shown in the Table on page 148.

Exercise No. 143

Mental Division

Divide mentally by 5 the answers to Exercise No. 50 as given on page 163. Compare your answers with Table I on page 7 .

Exercise No. 144

Continuous Addition Drill

Count by 11's to 275.
Count by 12's to 300.
Count by 13's to 325.
Count by 14's to 350.
Count by 15's to 375.
Count by 16's to 400.
Count by 17's to 425.
Count by 18's to 450.

Repeat this exercise three times.

Exercise No. 145

Multiplication Table Drill

Use Table II on page 48.

Multiply mentally the numbers in Column E by 9, 11, 12, 13, 14, 15, 16, 17 and 18.

Repeat this exercise three times.

Exercise No. 146
Factoring

Factor the numbers from 255 to 284 inclusive in the form shown in the table on page 148.

Exercise No. 147

Mental Division

Divide mentally by 6 the answers to Exercise No. 52 as given on page 163. Compare your answers with Table I on page 7.

Exercise No. 148

Mental Multiplication

Multiply mentally by 18 the numbers in Table I on page 7.

Exercise No. 149

Written Multiplication

Multiply by 1718 the numbers in Table III on page 49. Make a single multiplication of pairs of figures in the multiplicand up to 18.

Exercise No. 150

Factoring

Factor the numbers from 270 to 299 inclusive in the form shown in the table on pages 148.

Exercise No. 151

Mental Division

Divide mentally by 7 the answers to Exercise No. 53 as given on pages 163 and 164. Compare your answers with Table I on page 7.

Exercise No. 152

Continuous Addition Drill

Count by 12's to 300.
Count by 13's to 325.
Count by 14's to 350.
Count by 15's to 375.
Count by 16's to 400.
Count by 17's to 425.
Count by 18's to 450.
Count by 19's to 475.

Repeat this exercise three times.

Exercise No. 153

Multiplication Table Drill

Use Table II on page 48.

Multiply mentally the numbers in Column F by 11, 12, 13, 14, 15, 16, 17, 18 and 19.

Repeat this exercise three times.

Exercise No. 154

Factoring

Factor the numbers from 285 to 312 inclusive in the form shown in the table on page 148.

Exercise No. 155

Mental Division

Divide mentally by 8 the answers to Exercise No. 56 as given on page 164. Compare your answers with Table I on page 7 .

Exercise No. 156

Mental Multiplication

Multiply mentally by 19 the numbers in Table I on page 7 .

Exercise No. 157

Factoring

Factor the numbers from 300 to 328 inclusive in the form shown in the table on page 148.

Exercise No. 158

Mental Division

Divide mentally by 9 the answers to Exercise No. 60 as given on page 164. Compare your answers with Table I on page 7 .

Exercise No. 159

Written Multiplication

Multiply by 1819 the numbers in Table III on page 49. Make a single multiplication of pairs of figures in the multiplicand up to 19.

Exercise No. 160

Factoring

Factor the numbers from 313 to 343 inclusive in the form shown in the table on page 149.

Exercise No. 161

Mental Division

Divide mentally by 11 the answers to Exercise No. 61 as given on page 165. Compare your answers with Table I on page 7 .

Exercise No. 162

Multiplication Table Drill

Use Table II on page 48.
Multiply mentally the numbers in Column G by 12, 13, 14, 15, 16, 17, 18, 19 and 20.

Exercise No. 163

Factoring

Factor the numbers from 329 to 359 inclusive in the form shown in the table on pages 148 and 149.

Exercise No. 164

Mental Division

Divide mentally by 12 the answers to Exercise No. 77 as given on page 166. Compare your answers with Table I on page 7.

Exercise No. 165

Mental Multiplication

Multiply mentally by 20 the numbers in Table I on page 7.

Exercise No. 166

Written Multiplication

Multiply by 1920 the numbers in Table III on page 49. Make a single multiplication of pairs of figures in the multiplicand up to 20.

Exercise No. 167

Factoring

Factor the numbers from 344 to 372 inclusive in the form shown in the table on page 149.

Exercise No. 168

Mental Division

Divide mentally by 13 the answers to Exercise No. 90 as given on page 167. Compare your answers with Table I on page 7.

Exercise No. 169

Continuous Addition Drill

Count by 13's to 325.
Count by 14's to 350.
Count by 15's to 375.
Count by 16's to 400.
Count by 17's to 425.
Count by 18's to 450.
Count by 19's to 475.
Count by 21's to 525.

Exercise No. 170

Multiplication Table Drill

Use Table II on page 48.

Multiply mentally the numbers in Column H by 12, 13, 14, 15, 16, 17, 18, 19, 20 and 21.

Exercise No. 171

Factoring

Factor the numbers from 360 to 386 inclusive in the form shown in the table on page 149.

Exercise No. 172

Mental Multiplication

Multiply mentally by 21 the numbers in Table I on page 7.

Exercise No. 173

Written Multiplication

Multiply by 2021 the numbers in Table III on page 49. Make a single multiplication of pairs of figures in the multiplicand up to 21.

Exercise No. 174

Factoring

Factor the numbers from 373 to 399 inclusive in the form shown in the table on pages 149 and 150.

Exercise No. 175

Mental Division

Divide mentally by 14 the answers to Exercise No. 106 as given on page 168. Compare your answers with Table I on page 7.

Exercise No. 176

Continuous Addition Drill

Count by 14's to 350.
Count by 15's to 375.
Count by 16's to 400.
Count by 17's to 425.
Count by 18's to 450.
Count by 19's to 475.
Count by 21's to 525.
Count by 22's to 550.
Repeat this exercise three times.

Exercise No. 177

Multiplication Table Drill

Use Table II on page 48.
Multiply mentally the numbers in Column J by 13, 14, 15, 16, 17, 18, 19, 20, 21 and 22.

Exercise No. 178

Factoring

Factor the numbers from 387 to 413 inclusive in the form shown in the table on pages 149 and 150.

Exercise No. 179

Mental Multiplication

Multiply mentally by 22 the numbers in Table I on page 7 .

Exercise No. 180

Written Multiplication

Multiply by 2122 the numbers in Table III on page 49. Make a single multiplication of pairs of figures in the multiplicand up to 22.

Exercise No. 181

Factoring

Factor the numbers from 400 to 427 inclusive in the form shown in the table on page 150 .

Exercise No. 182

Mental Division

Divide mentally by 15 the answers to Exercise No. 119 as given on page 169. Compare your answers with Table I on page 7.

Exercise No. 183

Continuous Addition Drill

Count by 15's to 375.
Count by 16's to 400.
Count by 17's to 425.
Count by 18's to 450.
Count by 19's to 475.
Count by 21's to 525.
Count by 22's to 550.
Count by 23's to 575.

Repeat this exercise three times.

Exercise No. 184

Multiplication Table Drill

Use Table II on page 48.
Multiply mentally the numbers in Column K by 14, 15, 16, 17, 18, 19, 20, 21, 22 and 23.

Exercise No. 185

Factoring

Factor the numbers from 414 to 440 inclusive in the form shown in the table on page 150.

Exercise No. 186

Mental Multiplication

Multiply mentally by 23 the numbers in Table I on page 7.

Exercise No. 187

Written Multiplication

Multiply by 2223 the numbers in Table III on page 49. Make a single multiplication of pairs of figures in the multiplicand up to 23.

Exercise No. 188

Factoring

Factor the numbers from 428 to 455 inclusive in the form shown in the table on page 150.

Exercise No. 189

Mental Division

Divide mentally by 16 the answers to Exercise No. 131 as given on pages 169 and 170. Compare your answers with Table I on page 7 .

Exercise No. 190

Continuous Addition Drill

Count by 16's to 400.
Count by 17's to 425.
Count by 18's to 450.
Count by 19's to 475.
Count by 21's to 525.
Count by 22's to 550.
Count by 23's to 575.
Count by 24's to 600.

Repeat this exercise three times.

Exercise No. 191

Multiplication Table Drill

Use Table II on page 48.

Multiply mentally the numbers in Column L by 15, 16, 17, 18, 19, 20, 21, 22, 23 and 24.

Exercise No. 192

Factoring

Factor the numbers from 441 to 467 inclusive in the form shown in the table on pages 150 and 151.

Exercise No. 193

Mental Multiplication

Multiply mentally by 24 the numbers in Table I on page 7 .

Exercise No. 194

Written Multiplication

Multiply by 2324 the numbers in Table III on page 49. Make a single multiplication of pairs of figures in the multiplicand up to 24.

Exercise No. 195

Factoring

Factor the numbers from 456 to 479 inclusive in the form shown in the table on pages 150 and 151.

Exercise No. 196

Mental Division

Divide mentally by 17 the answers to Exercise No. 140 as given on page 170. Compare your answers with Table I on page 7 .

Exercise No. 197

Continuous Addition Drill

Count by 17's to 425.
Count by 18's to 450.
Count by 19's to 475.

Count by 21's to 525.
Count by 22's to 550.
Count by 23's to 575.
Count by 24's to 600.
Count by 25's to 625.
Repeat this exercise three times.

Exercise No. 198
Multiplication Table Drill

Use Table II on page 48.
Multiply mentally the numbers in Column M by 16, 17, 18, 19, 20, 21, 22, 23, 24 and 25.

Exercise No. 199
Factoring

Factor the numbers from 468 to 491 inclusive in the form shown in the table on page 151.

Exercise No. 200
Mental Multiplication

Multiply mentally by 25 the numbers in Table I on page 7.

Exercise No. 201
Written Multiplication

Multiply by 2425 the numbers in Table III on page 49. Make a single multiplication of pairs of figures in the multiplicand up to 25.

Exercise No. 202
Factoring

Factor the numbers from 480 to 503 inclusive in the form shown in the table on page 151.

Exercise No. 203
Mental Division
Divide mentally by 18 the answers to Exercise No. 148 as given on page 170 and 171. Compare your answers with Table I on page 7.

Exercise No. 204
Mental Multiplication
Multiply mentally by 20 the numbers in Table I on page 7.

Exercise No. 205
Continuous Addition Drill
Count by 18's to 450.
Count by 19's to 475.
Count by 21's to 525.
Count by 22's to 550.
Count by 23's to 575.
Count by 24's to 600.
Count by 25's to 625.
Repeat this exercise three times.

Exercise No. 206
Factoring
Factor the numbers from 492 to 515 inclusive in the form shown in the table on page 151.

Exercise No. 207
Continuous Addition Drill
Count by 19's to 475.
Count by 21's to 525.
Count by 22's to 550.
Count by 23's to 575.
Count by 24's to 600.
Count by 25's to 625.
Repeat this exercise three times.

Exercise No. 208
Mental Multiplication
Multiply mentally by 30 the numbers in Table I on page 7.

Exercise No. 209
Factoring
Factor the numbers from 504 to 527 inclusive in the form shown in the table on page 151.

Exercise No. 210
Mental Division
Divide mentally by 19 the answers to Exercise No. 149 as given on page 171. Compare your answers with Table I on page 7.

Exercise No. 211
Continuous Addition Drill
Count by 21's to 525.
Count by 22's to 550.
Count by 23's to 575.
Count by 24's to 600.
Count by 25's to 625.
Repeat this exercise three times.

Exercise No. 212
Mental Multiplication
Multiply mentally by 40 the numbers in Table I on page 7.

Exercise No. 213
Factoring
Factor the numbers from 516 to 539 inclusive in the form shown in the table on page 151.

Exercise No. 214
Continuous Addition Drill
Count by 22's to 550.
Count by 23's to 575.
Count by 24's to 600.
Count by 25's to 625.

Repeat this exercise three times.

Exercise No. 215
Mental Multiplication
Multiply mentally by 50 the numbers in Table I on page 7.

Exercise No. 216
Factoring
Factor the numbers from 528 to 551 inclusive in the form shown in the table on pages 151 and 152.

Exercise No. 217
Continuous Addition Drill
Count by 23's to 575.
Count by 24's to 600.
Count by 25's to 625.

Repeat this exercise three times.

Exercise No. 218
Mental Division
Divide mentally by 20 the answers to Exercise No. 165 as given on page 172. Compare your answers with Table I on page 7.

Exercise No. 219
Mental Multiplication
Multiply mentally by 60 the numbers in Table I on page 7.

Exercise No. 220
Factoring
Factor the numbers from 540 to 564 inclusive in the form shown in the table on page 152.

Exercise No. 221
Continuous Addition Drill
Count by 24's to 600.
Count by 25's to 625.
Repeat this exercise three times.

Exercise No. 222
Mental Multiplication
Multiply mentally by 70 the numbers in Table I on page 7.

Exercise No. 223
Factoring
Factor the numbers from 552 to 576 inclusive in the form shown in the table on page 152.

Exercise No. 224
Mental Division
Divide mentally by 21 the answers to Exercise No. 172 as given on page 172. Compare your answers with Table I on page 7.

Exercise No. 225

Continuous Addition Drill

Count by 25's to 625.

Repeat this exercise three times.

Exercise No. 226

Mental Multiplication

Multiply mentally by 80 the numbers in Table I on page 7 .

Exercise No. 227

Factoring

Factor the numbers from 565 to 592 inclusive in the form shown in the table on page 152.

Exercise No. 228

Mental Multiplication

Multiply mentally by 90 the numbers in Table I on page 7 .

Exercise No. 229

Multiplying Three Figures by One

We are now ready to start the mental multiplication of numbers of three places by numbers of one place. Work from left to right. Immediately name the first partial product as hundreds or thousands. Thus, taking the fourth example, this would be calculated as 800, 900, 902. The fifth example would be figured as 1000, 1120, 1124.

When dealing with numbers in the thousands be sure to consider the thousands as such and not as so many hundreds. If you wish, however, you may shorten the terminology. You may, for instance, think of one thousand one hundred twenty-six simply as one, one twenty-six, or as one, one two six.

Perform mentally the following multiplications.

1. 121 × 2	**8.** 842 × 2	**15.** 663 × 2
2. 232 × 2	**9.** 953 × 2	**16.** 721 × 2
3. 343 × 2	**10.** 161 × 2	**17.** 832 × 2
4. 451 × 2	**11.** 222 × 2	**18.** 943 × 2
5. 562 × 2	**12.** 333 × 2	**19.** 151 × 2
6. 623 × 2	**13.** 441 × 2	**20.** 262 × 2
7. 731 × 2	**14.** 552 × 2	

Exercise No. 230

Factoring

Factor the numbers from 577 to 605 inclusive in the form shown in the table on page 152.

Exercise No. 231

Mental Division

Divide mentally by 22 the answers to Exercise No. 179 as given on page 173. Compare your answers with Table I on page 7 .

Exercise No. 232

Mental Multiplication

Multiply mentally by 110 the numbers in Table I on page 7 .

Exercise No. 233

Multiplying Three Figures by One

Perform mentally the following multiplications.

1. 131 × 3	**3.** 353 × 3	**5.** 571 × 3
2. 242 × 3	**4.** 464 × 3	**6.** 632 × 3

7. 743 × 3	12. 344 × 3	17. 841 × 3
8. 854 × 3	13. 451 × 3	18. 952 × 3
9. 961 × 3	14. 562 × 3	19. 163 × 3
10. 172 × 3	15. 673 × 3	20. 274 × 3
11. 233 × 3	16. 734 × 3	

Exercise No. 234
Factoring

Factor the numbers from 593 to 625 inclusive in the form shown in the table on pages 152 and 153.

Exercise No. 235
Mental Division

Divide mentally by 23 the answers to Exercise No. 186 as given on pages 173 and 174. Compare your answers with Table I on page 7.

Exercise No. 236
Mental Multiplication

Multiply mentally by 120 the numbers in Table I on page 7.

Exercise No. 237
Multiplying Three Figures by One

Perform mentally the following multiplications.

1. 141 × 4	8. 863 × 4	15. 685 × 4
2. 252 × 4	9. 974 × 4	16. 741 × 4
3. 363 × 4	10. 185 × 4	17. 852 × 4
4. 474 × 4	11. 241 × 4	18. 963 × 4
5. 585 × 4	12. 352 × 4	19. 174 × 4
6. 641 × 4	13. 463 × 4	20. 285 × 4
7. 752 × 4	14. 574 × 4	

Exercise No. 238
Mental Division
Divide mentally by 24 the answers to Exercise No. 193 as given on page 174. Compare your answers with Table I on page 7.

Exercise No. 239
Mental Multiplication
Multiply mentally by 130 the numbers in Table I on page 7.

Exercise No. 240
Multiplying Three Figures by One
Perform mentally the following multiplications.

1. 151 × 5	8. 872 × 5	15. 693 × 5
2. 262 × 5	9. 983 × 5	16. 754 × 5
3. 373 × 5	10. 194 × 5	17. 865 × 5
4. 484 × 5	11. 255 × 5	18. 976 × 5
5. 595 × 5	12. 366 × 5	19. 181 × 5
6. 656 × 5	13. 471 × 5	20. 292 × 5
7. 761 × 5	14. 582 × 5	

Exercise No. 241
Mental Division
Divide mentally by 25 the answers to Exercise No. 200 as given on pages 174 and 175. Compare your answers with Table I on page 7.

Exercise No. 242
Mental Multiplication
Multiply mentally by 140 the numbers in Table I on page 7.

Exercise No. 243
Multiplying Three Figures by One
Perform mentally the following multiplications.

1. 141 × 6	8. 851 × 6	15. 661 × 6
2. 252 × 6	9. 962 × 6	16. 772 × 6
3. 363 × 6	10. 173 × 6	17. 883 × 6
4. 474 × 6	11. 284 × 6	18. 994 × 6
5. 585 × 6	12. 395 × 6	19. 145 × 6
6. 696 × 6	13. 446 × 6	20. 256 × 6
7. 747 × 6	14. 557 × 6	

Exercise No. 244
Mental Multiplication
Multiply mentally by 150 the numbers in Table I on page 7 .

Exercise No. 245
Multiplying Three Figures by One
Perform mentally the following multiplications.

1. 131 × 7	8. 838 × 7	15. 637 × 7
2. 242 × 7	9. 941 × 7	16. 748 × 7
3. 353 × 7	10. 152 × 7	17. 851 × 7
4. 464 × 7	11. 263 × 7	18. 962 × 7
5. 575 × 7	12. 374 × 7	19. 173 × 7
6. 686 × 7	13. 485 × 7	20. 284 × 7
7. 797 × 7	14. 596 × 7	

Exercise No. 246
Mental Multiplication
Multiply mentally by 160 the numbers in Table I on page 7 .

Exercise No. 247
Multiplying Three Figures by One
Perform mentally the following multiplications.

1. 141 × 8	**8.** 858 × 8	**15.** 666 × 8
2. 252 × 8	**9.** 969 × 8	**16.** 777 × 8
3. 363 × 8	**10.** 171 × 8	**17.** 888 × 8
4. 474 × 8	**11.** 282 × 8	**18.** 999 × 8
5. 585 × 8	**12.** 393 × 8	**19.** 741 × 8
6. 696 × 8	**13.** 444 × 8	**20.** 652 × 8
7. 747 × 8	**14.** 555 × 8	

FRACTIONS IN GENERAL

The multiplication or the division of fractions will present no difficulty to the student of these pages since it is simply a matter of combining operations in which he is well practised. What needs more particular attention is the addition and subtraction of the kinds of fractions most commonly encountered in practical work in office, shop and home. The average person would immediately reach for a pencil if asked the sum of $\frac{3}{4}$ and $\frac{5}{8}$ or the difference between $1\frac{1}{3}$ and $\frac{3}{8}$. Yet a little practice with calculations of this kind makes it very easy to perform them mentally.

The succeeding examples in addition and subtraction of fractions are based on the possible combinations of two fractions of the orders of halves, quarters, eighths, sixteenths, thirds, sixths, twelfths, fifths and tenths.

These exercises are to stimulate memory and rapid thinking. No instructions are given as to how to perform them because it is assumed that the student is familiar with the reduction of fractions to a common denominator.

Exercise No. 248
Reduction of Fractions

1. Reduce to eighths: $\frac{1}{2}$, $\frac{1}{4}$, $\frac{3}{4}$

2. Reduce to sixteenths: $\frac{1}{8}$, $\frac{1}{4}$, $\frac{3}{8}$, $\frac{1}{2}$, $\frac{5}{8}$, $\frac{3}{4}$, $\frac{7}{8}$

3. Reduce to sixths: $\frac{1}{3}$, $\frac{1}{2}$, $\frac{2}{3}$

4. Reduce to twelfths: $\frac{1}{6}$, $\frac{1}{4}$, $\frac{1}{3}$, $\frac{1}{2}$, $\frac{2}{3}$, $\frac{3}{4}$, $\frac{5}{6}$

5. Reduce to twenty-fourths: $\frac{1}{12}$, $\frac{1}{8}$, $\frac{1}{6}$, $\frac{1}{4}$, $\frac{1}{3}$, $\frac{5}{12}$, $\frac{1}{2}$, $\frac{7}{12}$, $\frac{5}{8}$, $\frac{2}{3}$, $\frac{3}{4}$, $\frac{5}{6}$, $\frac{11}{12}$

6. Reduce to tenths: $\frac{1}{5}$, $\frac{2}{5}$, $\frac{1}{2}$, $\frac{3}{5}$, $\frac{4}{5}$

7. Reduce to twentieths: $\frac{1}{10}$, $\frac{1}{5}$, $\frac{3}{10}$, $\frac{2}{5}$, $\frac{1}{2}$, $\frac{3}{5}$, $\frac{7}{10}$, $\frac{4}{5}$, $\frac{9}{10}$

8. Reduce to fortieths: $\frac{1}{10}$, $\frac{1}{8}$, $\frac{1}{5}$, $\frac{1}{4}$, $\frac{3}{10}$, $\frac{3}{8}$, $\frac{2}{5}$, $\frac{1}{2}$, $\frac{3}{5}$, $\frac{5}{8}$, $\frac{7}{10}$, $\frac{3}{4}$, $\frac{4}{5}$, $\frac{7}{8}$, $\frac{9}{10}$

9. Reduce to fifteenths: $\frac{1}{5}$, $\frac{1}{3}$, $\frac{2}{5}$, $\frac{3}{5}$, $\frac{2}{3}$, $\frac{4}{5}$

10. Reduce to thirtieths: $\frac{1}{10}$, $\frac{1}{6}$, $\frac{1}{5}$, $\frac{3}{10}$, $\frac{1}{3}$, $\frac{2}{5}$, $\frac{1}{2}$, $\frac{3}{5}$, $\frac{2}{3}$, $\frac{7}{10}$, $\frac{4}{5}$, $\frac{5}{6}$, $\frac{9}{10}$

Exercise No. 249
Mental Multiplication
Multiply mentally by 170 the numbers in Table I on page **7**.

Exercise No. 250
Addition of Fractions
Add the following mentally.

1. $\frac{1}{2} + \frac{1}{4}$	**11.** $\frac{3}{4} + \frac{1}{8}$	**21.** $\frac{1}{2} + \frac{13}{16}$	**31.** $\frac{3}{4} + \frac{1}{16}$
2. $\frac{1}{2} + \frac{3}{4}$	**12.** $\frac{3}{4} + \frac{3}{8}$	**22.** $\frac{1}{2} + \frac{15}{16}$	**32.** $\frac{3}{4} + \frac{3}{16}$
3. $\frac{1}{2} + \frac{1}{8}$	**13.** $\frac{3}{4} + \frac{5}{8}$	**23.** $\frac{1}{4} + \frac{1}{16}$	**33.** $\frac{3}{4} + \frac{5}{16}$
4. $\frac{1}{2} + \frac{3}{8}$	**14.** $\frac{3}{4} + \frac{7}{8}$	**24.** $\frac{1}{4} + \frac{3}{16}$	**34.** $\frac{3}{4} + \frac{7}{16}$
5. $\frac{1}{2} + \frac{5}{8}$	**15.** $\frac{1}{2} + \frac{1}{16}$	**25.** $\frac{1}{4} + \frac{5}{16}$	**35.** $\frac{3}{4} + \frac{9}{16}$
6. $\frac{1}{2} + \frac{7}{8}$	**16.** $\frac{1}{2} + \frac{3}{16}$	**26.** $\frac{1}{4} + \frac{7}{16}$	**36.** $\frac{3}{4} + \frac{11}{16}$
7. $\frac{1}{4} + \frac{1}{8}$	**17.** $\frac{1}{2} + \frac{5}{16}$	**27.** $\frac{1}{4} + \frac{9}{16}$	**37.** $\frac{3}{4} + \frac{13}{16}$
8. $\frac{1}{4} + \frac{3}{8}$	**18.** $\frac{1}{2} + \frac{7}{16}$	**28.** $\frac{1}{4} + \frac{11}{16}$	**38.** $\frac{3}{4} + \frac{15}{16}$
9. $\frac{1}{4} + \frac{5}{8}$	**19.** $\frac{1}{2} + \frac{9}{16}$	**29.** $\frac{1}{4} + \frac{13}{16}$	**39.** $\frac{1}{8} + \frac{1}{16}$
10. $\frac{1}{4} + \frac{7}{8}$	**20.** $\frac{1}{2} + \frac{11}{16}$	**30.** $\frac{1}{4} + \frac{15}{16}$	**40.** $\frac{1}{8} + \frac{3}{16}$

Exercise No. 251
Multiplying Three Figures by One

1. 152×9	**8.** 869×9	**15.** 679×9
2. 263×9	**9.** 973×9	**16.** 784×9
3. 374×9	**10.** 184×9	**17.** 895×9
4. 485×9	**11.** 295×9	**18.** 946×9
5. 596×9	**12.** 346×9	**19.** 157×9
6. 647×9	**13.** 457×9	**20.** 268×9
7. 758×9	**14.** 568×9	

Exercise No. 252

Mental Division

Divide mentally by 2 the answers to Exercise No. 229 as given on page 175.

Exercise No. 253

Addition of Fractions

Do the last thirty examples in Exercise No. 250 on the preceding page, and also add the following.

1. $\frac{1}{8} + \frac{5}{16}$ **4.** $\frac{1}{8} + \frac{11}{16}$ **7.** $\frac{3}{8} + \frac{1}{16}$ **10.** $\frac{3}{8} + \frac{7}{16}$

2. $\frac{1}{8} + \frac{7}{16}$ **5.** $\frac{1}{8} + \frac{13}{16}$ **8.** $\frac{3}{8} + \frac{3}{16}$

3. $\frac{1}{8} + \frac{9}{16}$ **6.** $\frac{1}{8} + \frac{15}{16}$ **9.** $\frac{3}{8} + \frac{5}{16}$

Exercise No. 254

Mental Multiplication

Multiply mentally by 180 the numbers in Table I on page 7.

Exercise No. 255

Mental Division

Divide mentally by 3 the answers to Exercise No. 233 as given on page 175. Compare your answers with Exercise No. 233.

Exercise No. 256

Addition of Fractions

Review the last twenty examples in Exercise No. 250 on page 97 and those in Exercise No. 253 on page 98. Also add the following.

1. $\frac{3}{8} + \frac{9}{16}$ **4.** $\frac{3}{8} + \frac{15}{16}$ **7.** $\frac{5}{8} + \frac{5}{16}$ **10.** $\frac{5}{8} + \frac{11}{16}$

2. $\frac{3}{8} + \frac{11}{16}$ **5.** $\frac{5}{8} + \frac{1}{16}$ **8.** $\frac{5}{8} + \frac{7}{16}$

3. $\frac{3}{8} + \frac{13}{16}$ **6.** $\frac{5}{8} + \frac{5}{16}$ **9.** $\frac{5}{8} + \frac{9}{16}$

Exercise No. 257
Mental Multiplication

Multiply mentally by 190 the numbers in Table I on page 7 .

Exercise No. 258
Mental Division

Divide mentally by 4 the answers to Exercise No. 237 as given on page 175.

Exercise No. 259
Addition of Fractions

Review the last ten examples in Exercise No. 250 on page 97 , as well as those in Exercise No. 253 on page 98 and Exercise No. 256 on page 98 . Also add the following.

1. $\frac{5}{8} + \frac{13}{16}$ 4. $\frac{7}{8} + \frac{3}{16}$ 7. $\frac{7}{8} + \frac{9}{16}$ 10. $\frac{7}{8} + \frac{15}{16}$

2. $\frac{5}{8} + \frac{15}{16}$ 5. $\frac{7}{8} + \frac{5}{16}$ 8. $\frac{7}{8} + \frac{11}{16}$

3. $\frac{7}{8} + \frac{1}{16}$ 6. $\frac{7}{8} + \frac{7}{16}$ 9. $\frac{7}{8} + \frac{13}{16}$

Exercise No. 260
Mental Multiplication

Multiply mentally by 200 the numbers in Table I on page 7 .

Exercise No. 261
Addition of Fractions

Review the examples in Exercise No. 253 on page 98 , No. 256 on page 98 and No. 259 above. Also add the following.

1. $\frac{1}{3} + \frac{1}{6}$ 4. $\frac{1}{3} + \frac{5}{12}$ 7. $\frac{2}{3} + \frac{1}{12}$ 10. $\frac{2}{3} + \frac{11}{12}$

2. $\frac{2}{3} + \frac{1}{6}$ 5. $\frac{1}{3} + \frac{7}{12}$ 8. $\frac{2}{3} + \frac{5}{12}$

3. $\frac{1}{3} + \frac{1}{12}$ 6. $\frac{1}{3} + \frac{11}{12}$ 9. $\frac{2}{3} + \frac{7}{12}$

Exercise No. 262
Mental Division

Divide mentally by 5 the answers to Exercise No. 240 as given on page 175.

Exercise No. 263
Subtraction of Fractions

Perform mentally the following subtractions.

1. $\frac{3}{4} - \frac{1}{2}$	8. $\frac{5}{8} - \frac{1}{4}$	16. $\frac{11}{16} - \frac{1}{2}$	24. $\frac{7}{16} - \frac{1}{4}$
2. $1\frac{1}{4} - \frac{1}{2}$	9. $\frac{7}{8} - \frac{1}{4}$	17. $\frac{13}{16} - \frac{1}{2}$	25. $\frac{9}{16} - \frac{1}{4}$
3. $\frac{5}{8} - \frac{1}{2}$	10. $1\frac{1}{8} - \frac{1}{4}$	18. $\frac{15}{16} - \frac{1}{2}$	26. $\frac{11}{16} - \frac{1}{4}$
4. $\frac{7}{8} - \frac{1}{2}$	11. $\frac{7}{8} - \frac{3}{4}$	19. $1\frac{1}{16} - \frac{1}{2}$	27. $\frac{13}{16} - \frac{1}{4}$
5. $1\frac{1}{8} - \frac{1}{2}$	12. $1\frac{1}{8} - \frac{3}{4}$	20. $1\frac{3}{16} - \frac{1}{2}$	28. $\frac{15}{16} - \frac{1}{4}$
6. $1\frac{3}{8} - \frac{1}{2}$	13. $1\frac{3}{8} - \frac{3}{4}$	21. $1\frac{5}{16} - \frac{1}{2}$	29. $1\frac{1}{16} - \frac{1}{4}$
7. $\frac{3}{8} - \frac{1}{4}$	14. $1\frac{5}{8} - \frac{3}{4}$	22. $1\frac{7}{16} - \frac{1}{2}$	30. $1\frac{3}{16} - \frac{1}{4}$
	15. $\frac{9}{16} - \frac{1}{2}$	23. $\frac{5}{16} - \frac{1}{4}$	

Exercise No. 264
Mental Multiplication

Multiply mentally by 210 the numbers in Table I on page 7 .

Exercise No. 265
Subtraction of Fractions

Review the last twenty examples in Exercise No. 263 above, and also perform the following subtractions.

1. $\frac{13}{16} - \frac{3}{4}$	4. $1\frac{3}{16} - \frac{3}{4}$	7. $1\frac{9}{16} - \frac{3}{4}$	10. $\frac{5}{16} - \frac{1}{8}$
2. $\frac{15}{16} - \frac{3}{4}$	5. $1\frac{5}{16} - \frac{3}{4}$	8. $1\frac{11}{16} - \frac{3}{4}$	
3. $1\frac{1}{16} - \frac{3}{4}$	6. $1\frac{7}{16} - \frac{3}{4}$	9. $\frac{3}{16} - \frac{1}{8}$	

Exercise No. 266
Mental Division

Divide mentally by 6 the answers to Exercise No. 243 as given on page 175.

Exercise No. 267
Addition of Fractions

Review the examples in Exercise No. 256 on page 98 , No. 259 on page 99 and No. 261 on page 99 . Also perform the following additions.

1. $\frac{1}{6} + \frac{1}{12}$ 4. $\frac{1}{6} + \frac{11}{12}$ 7. $\frac{5}{6} + \frac{7}{12}$ 10. $\frac{1}{2} + \frac{2}{3}$

2. $\frac{1}{6} + \frac{5}{12}$ 5. $\frac{5}{6} + \frac{1}{12}$ 8. $\frac{5}{6} + \frac{11}{12}$

3. $\frac{1}{6} + \frac{7}{12}$ 6. $\frac{5}{6} + \frac{5}{12}$ 9. $\frac{1}{2} + \frac{1}{3}$

Exercise No. 268
Mental Multiplication

Multiply mentally by 220 the numbers in Table I on page 7 .

Exercise No. 269
Subtraction of Fractions

Review the last ten examples in Exercise No. 263 on page 100 and No. 265 on page 100. Also perform the following subtractions.

1. $\frac{7}{16} - \frac{1}{8}$ 4. $\frac{13}{16} - \frac{1}{8}$ 7. $\frac{7}{16} - \frac{3}{8}$ 10. $\frac{13}{16} - \frac{3}{8}$

2. $\frac{9}{16} - \frac{1}{8}$ 5. $\frac{15}{16} - \frac{1}{8}$ 8. $\frac{9}{16} - \frac{3}{8}$

3. $\frac{11}{16} - \frac{1}{8}$ 6. $1\frac{1}{16} - \frac{1}{8}$ 9. $\frac{11}{16} - \frac{3}{8}$

Exercise No. 270
Mental Division

Divide mentally by 7 the answers to Exercise No. 245 as given on page 176.

Exercise No. 271

Addition of Fractions

Review the examples in Exercise No. 259 on page 99 , No. 261 on page 99 and No. 267 on page 101. Also perform the following additions.

1. $\frac{1}{2} + \frac{1}{6}$ **4.** $\frac{1}{4} + \frac{5}{6}$ **7.** $\frac{1}{8} + \frac{1}{6}$ **10.** $\frac{7}{8} + \frac{1}{6}$

2. $\frac{1}{2} + \frac{5}{6}$ **5.** $\frac{3}{4} + \frac{1}{6}$ **8.** $\frac{3}{8} + \frac{1}{6}$

3. $\frac{1}{4} + \frac{1}{6}$ **6.** $\frac{3}{4} + \frac{5}{6}$ **9.** $\frac{5}{8} + \frac{1}{6}$

Exercise No. 272

Mental Multiplication

Multiply mentally by 230 the numbers in Table I on page 7 .

Exercise No. 273

Subtraction of Fractions

Review the examples in Exercise No. 265 on page 100 and No. 269 on page 101. Also perform the following subtractions.

1. $\frac{15}{16} - \frac{3}{8}$ **4.** $1\frac{5}{16} - \frac{3}{8}$ **7.** $\frac{15}{16} - \frac{5}{8}$ **10.** $1\frac{5}{16} - \frac{5}{8}$

2. $1\frac{1}{16} - \frac{3}{8}$ **5.** $\frac{11}{16} - \frac{5}{8}$ **8.** $1\frac{1}{16} - \frac{5}{8}$

3. $1\frac{3}{16} - \frac{3}{8}$ **6.** $\frac{13}{16} - \frac{5}{8}$ **9.** $1\frac{3}{16} - \frac{5}{8}$

Exercise No. 274

Mental Division

Divide mentally by 8 the answers to Exercise No. 247 as given on page 176.

Exercise No. 275

Addition of Fractions

Review the examples in Exercise No. 261 on page 99 , No. 267 on page 101 and No. 271 on this page. Also perform the following additions.

1. $\frac{1}{8} + \frac{5}{6}$ 4. $\frac{7}{8} + \frac{5}{6}$ 7. $\frac{1}{2} + \frac{7}{12}$ 10. $\frac{1}{4} + \frac{5}{12}$

2. $\frac{3}{8} + \frac{5}{6}$ 5. $\frac{1}{2} + \frac{1}{12}$ 8. $\frac{1}{2} + \frac{11}{12}$

3. $\frac{5}{8} + \frac{5}{6}$ 6. $\frac{1}{2} + \frac{5}{12}$ 9. $\frac{1}{4} + \frac{1}{12}$

Exercise No. 276
Mental Multiplication

Multiply mentally by 240 the numbers in Table I on page 7.

Exercise No. 277
Subtraction of Fractions

Review the examples in Exercise No. 269 on page 101 and No. 273 on page 102. Also perform the following.

1. $1\frac{7}{16} - \frac{5}{8}$ 4. $1\frac{1}{16} - \frac{7}{8}$ 7. $1\frac{7}{16} - \frac{7}{8}$ 10. $1\frac{13}{16} - \frac{7}{8}$

2. $1\frac{9}{16} - \frac{5}{8}$ 5. $1\frac{3}{16} - \frac{7}{8}$ 8. $1\frac{9}{16} - \frac{7}{8}$

3. $\frac{15}{16} - \frac{7}{8}$ 6. $1\frac{5}{16} - \frac{7}{8}$ 9. $1\frac{11}{16} - \frac{7}{8}$

Exercise No. 278
Mental Division

Divide mentally by 9 the answers to Exercise No. 251 as given on page 176.

Exercise No. 279
Addition of Fractions

Review the examples in Exercise No. 267 on page 101, No. 271 on page 102 and No. 275 on this page. Also perform the following additions.

1. $\frac{1}{4} + \frac{7}{12}$ 4. $\frac{3}{4} + \frac{5}{12}$ 7. $\frac{1}{8} + \frac{1}{12}$ 10. $\frac{1}{8} + \frac{11}{12}$

2. $\frac{1}{4} + \frac{11}{12}$ 5. $\frac{3}{4} + \frac{7}{12}$ 8. $\frac{1}{8} + \frac{5}{12}$

3. $\frac{3}{4} + \frac{1}{12}$ 6. $\frac{3}{4} + \frac{11}{12}$ 9. $\frac{1}{8} + \frac{7}{12}$

Exercise No. 280
Mental Multiplication
Multiply mentally by 250 the numbers in Table I on page 7.

Exercise No. 281
Subtraction of Fractions
Review the examples in Exercise No. 273 on page 102 and No. 277 on page 103. Also perform the following subtractions.

1. $\frac{1}{2} - \frac{1}{3}$ 4. $\frac{3}{4} - \frac{1}{3}$ 7. $\frac{3}{4} - \frac{2}{3}$ 10. $1\frac{7}{12} - \frac{2}{3}$
2. $\frac{5}{6} - \frac{2}{3}$ 5. $\frac{11}{12} - \frac{1}{3}$ 8. $1\frac{1}{12} - \frac{2}{3}$
3. $\frac{5}{12} - \frac{1}{3}$ 6. $1\frac{1}{4} - \frac{1}{3}$ 9. $1\frac{1}{4} - \frac{2}{3}$

Exercise No. 282
Mental Division
Divide mentally the following. Express remainders as such instead of as fractions.

1. $328 \div 121$ 8. $1786 \div 842$ 15. $1998 \div 571$
2. $593 \div 232$ 9. $2114 \div 953$ 16. $690 \div 141$
3. $794 \div 343$ 10. $439 \div 161$ 17. $1208 \div 252$
4. $1249 \div 451$ 11. $406 \div 131$ 18. $1704 \div 363$
5. $1580 \div 562$ 12. $776 \div 242$ 19. $2178 \div 474$
6. $1835 \div 623$ 13. $1164 \div 353$ 20. $2620 \div 585$
7. $1774 \div 731$ 14. $1574 \div 464$

Exercise No. 283
Addition of Fractions
Review the examples in Exercise No. 271 on page 102, No. 275 on page 103 and No. 279 on page 103. Also perform the following additions.

1. $\frac{3}{8} + \frac{1}{12}$ 4. $\frac{3}{8} + \frac{11}{12}$ 7. $\frac{5}{8} + \frac{7}{12}$ 10. $\frac{7}{8} + \frac{5}{12}$
2. $\frac{3}{8} + \frac{5}{12}$ 5. $\frac{5}{8} + \frac{1}{12}$ 8. $\frac{5}{8} + \frac{11}{12}$
3. $\frac{3}{8} + \frac{7}{12}$ 6. $\frac{5}{8} + \frac{5}{12}$ 9. $\frac{7}{8} + \frac{1\frac{1}{12}}$

Exercise No. 284

Multiplying Two Figures by Two

With this exercise we start the general multiplication of two numbers of two places each. You have had some experience with such numbers in using the numbers up to 25 as direct multipliers. In the succeeding exercises, however, the multipliers are greater than 25 and the operation is performed differently.

Multiply the whole of the multiplicand by the first figure of the multiplier; next multiply the whole of the multiplicand by the second figure of the multiplier; and finally add the two partial products.

When you multiply the first figure of the multiplicand by the first figure of the multiplier you will get a number of either three places, as in the first example (where 20×40 produces 800), or four places, as in the second example (where 2×5 produces 10). Add to this first result as you work along from left to right. Similarly, when you multiply the first figure of the multiplicand by the second figure of the multiplier, you will get a number of either two or three places.

Repeat to yourself the original example and the partial products as often as you find necessary. The need for such repetitions will grow less as you become more practised.

Taking the first example: repeat, 41×26, 41×26, 41×26. 40×20 is 800, 1×2 is 2, 820. (say 1×2 rather than 1×20 because the former method is simpler when dealing with large numbers. When you think of the 2 as following the 8 it of course becomes a 20 in the product.) Repeat 820, 820, 820. 40×6 is 240, 1×6 is 6, 246. Repeat $820 + 246$, $820 + 246$, $820 + 246$. Add: 1020, 1060, 1066.

The second example is performed: 1000, 1020; 350, 357. $1020 + 357$, 1320, 1370, 1377.

Most of the examples in this exercise are very simple and there can be no objection to your shortening the method given, which is a general method applicable to increasingly larger numbers. Thus in the examples illustrated you should be able to note at a glance that the first partial products are 820 and 1020.

1. 41 × 26	8. 41 × 34	15. 41 × 33
2. 51 × 27	9. 51 × 26	16. 51 × 34
3. 61 × 28	10. 61 × 27	17. 61 × 26
4. 71 × 29	11. 71 × 28	18. 71 × 27
5. 8] × 31	12. 81 × 29	19. 81 × 28
6. 91 × 32	13. 91 × 31	20. 91 × 29
7. 31 × 33	14. 31 × 32	

Exercise No. 285

Subtraction of Fractions

Review the examples in Exercise No. 277 on page 103 and No. 281 on page 104. Also perform the following subtractions.

1. $\frac{1}{4} - \frac{1}{6}$	4. $1\frac{1}{12} - \frac{1}{6}$	7. $1\frac{5}{12} - \frac{5}{6}$	10. $1\frac{1}{6} - \frac{1}{2}$
2. $\frac{7}{12} - \frac{1}{6}$	5. $\frac{11}{12} - \frac{5}{6}$	8. $1\frac{3}{4} - \frac{5}{6}$	
3. $\frac{3}{4} - \frac{1}{6}$	6. $1\frac{1}{4} - \frac{5}{6}$	9. $\frac{5}{6} - \frac{1}{2}$	

Exercise No. 286

Mental Division

Divide mentally the following.

1. 445 ÷ 222	6. 2274 ÷ 632	11. 2830 ÷ 641
2. 695 ÷ 333	7. 2747 ÷ 743	12. 3233 ÷ 752
3. 1258 ÷ 441	8. 3242 ÷ 854	13. 3624 ÷ 863
4. 1655 ÷ 552	9. 3747 ÷ 961	14. 3989 ÷ 974
5. 1700 ÷ 663	10. 533 ÷ 172	15. 902 ÷ 185

16. $845 \div 151$ 18. $2013 \div 373$ 20. $3094 \div 595$
17. $1440 \div 262$ 19. $2564 \div 484$

Exercise No. 287
Addition of Fractions

Review the examples in Exercise No. 275 on page 103, No. 279 on page 103 and No. 283 on page 104. Also perform the following additions.

1. $\frac{7}{8} + \frac{7}{12}$ 4. $\frac{1}{5} + \frac{3}{10}$ 7. $\frac{2}{5} + \frac{1}{10}$ 10. $\frac{2}{5} + \frac{9}{10}$
2. $\frac{7}{8} + \frac{11}{12}$ 5. $\frac{1}{5} + \frac{7}{10}$ 8. $\frac{2}{5} + \frac{3}{10}$
3. $\frac{1}{5} + \frac{1}{10}$ 6. $\frac{1}{5} + \frac{9}{10}$ 9. $\frac{2}{5} + \frac{7}{10}$

Exercise No. 288
Multiplying Two Figures by Two

In doing exercises of this type always use the second number as the multiplier. Using the first example to illustrate, find 30 times 42 and then 5 times 42; do not work the other way around by finding 40 times 35 and then 2 times 35. This caution is given because of the special way in which the exercises are graded.

1. 42×35 8. 42×43 15. 42×42
2. 52×36 9. 52×35 16. 52×43
3. 62×37 10. 62×36 17. 62×34
4. 72×38 11. 72×37 18. 72×35
5. 82×39 12. 82×38 19. 82×36
6. 92×41 13. 92×39 20. 92×37
7. 32×42 14. 32×41

Exercise No. 289
Subtraction of Fractions

Review the examples in Exercise No. 277 on page 103 and No. 281 on page 104. Also perform the following subtractions.

1. $\frac{2}{3} - \frac{1}{2}$ 4. $1\frac{1}{24} - \frac{1}{4}$ 7. $\frac{7}{24} - \frac{1}{8}$ 10. $1\frac{1}{24} - \frac{7}{8}$

2. $1\frac{1}{3} - \frac{1}{2}$ 5. $\frac{11}{12} - \frac{3}{4}$ 8. $\frac{13}{24} - \frac{3}{8}$

3. $\frac{5}{12} - \frac{1}{4}$ 6. $1\frac{7}{12} - \frac{3}{4}$ 9. $\frac{19}{24} - \frac{5}{8}$

Exercise No. 290
Mental Division

1. $1479 \div 721$ 8. $1523 \div 451$ 15. $3012 \div 685$

2. $2435 \div 832$ 9. $1966 \div 562$ 16. $3347 \div 656$

3. $2036 \div 943$ 10. $2421 \div 673$ 17. $4498 \div 761$

4. $387 \div 151$ 11. $1156 \div 241$ 18. $4924 \div 872$

5. $623 \div 262$ 12. $1643 \div 352$ 19. $5547 \div 983$

6. $745 \div 233$ 13. $2128 \div 463$ 20. $1067 \div 194$

7. $1134 \div 344$ 14. $2581 \div 574$

Exercise No. 291

Addition of Fractions

Review the examples in Exercise No. 279 on page 103, No. 283 on page 104 and No. 287 on page 107. Also perform the following additions.

1. $\frac{3}{5} + \frac{1}{10}$ 4. $\frac{3}{5} + \frac{9}{10}$ 7. $\frac{4}{5} + \frac{7}{10}$ 10. $\frac{1}{2} + \frac{2}{5}$

2. $\frac{3}{5} + \frac{3}{10}$ 5. $\frac{4}{5} + \frac{1}{10}$ 8. $\frac{4}{5} + \frac{9}{10}$

3. $\frac{3}{5} + \frac{7}{10}$ 6. $\frac{4}{5} + \frac{3}{10}$ 9. $\frac{1}{2} + \frac{1}{5}$

Exercise No. 292

Mental Multiplication

Multiply mentally the following.

1. 43×44 8. 43×52 15. 43×51

2. 53×45 9. 53×44 16. 53×52

3. 63×46 10. 63×45 17. 63×44

4. 73×47 11. 73×46 18. 78×45

5. 83×48 12. 83×47 19. 83×46

6. 93×49 13. 93×48 20. 93×47

7. 33×51 14. 33×49

Exercise No. 293
Subtraction of Fractions

Review the examples in Exercise No. 281 on page 104 and No. 289 on page 108. Also do the following.

1. $\frac{23}{24} - \frac{1}{8}$ 4. $1\frac{17}{24} - \frac{7}{8}$ 7. $1\frac{1}{12} - \frac{1}{2}$ 10. $\frac{2}{3} - \frac{1}{4}$

2. $1\frac{5}{24} - \frac{3}{8}$ 5. $\frac{7}{12} - \frac{1}{2}$ 8. $1\frac{5}{12} - \frac{1}{2}$

3. $1\frac{11}{24} - \frac{5}{8}$ 6. $\frac{11}{12} - \frac{1}{2}$ 9. $\frac{1}{3} - \frac{1}{4}$

Exercise No. 294
Mental Division

Divide mentally the following.

1. $444 \div 131$ 8. $4716 \div 963$ 15. $3573 \div 693$

2. $795 \div 242$ 9. $815 \div 174$ 16. $971 \div 141$

3. $1154 \div 353$ 10. $1348 \div 285$ 17. $1712 \div 252$

4. $1424 \div 464$ 11. $1421 \div 255$ 18. $2255 \div 363$

5. $1767 \div 571$ 12. $2118 \div 366$ 19. $2955 \div 474$

6. $3186 \div 740$ 13. $2676 \div 471$ 20. $3820 \div 585$

7. $3493 \div 852$ 14. $3375 \div 582$

Exercise No. 295
Addition of Fractions

Review the examples in Exercise No. 279 on page 103, No. 283 on page 104 and No. 292 on page 108. Also perform the following additions.

1. $\frac{1}{2} + \frac{3}{5}$ 4. $\frac{1}{2} + \frac{3}{10}$ 7. $\frac{1}{4} + \frac{1}{5}$ 10. $\frac{1}{4} + \frac{4}{5}$

2. $\frac{1}{2} + \frac{4}{5}$ 5. $\frac{1}{2} + \frac{7}{10}$ 8. $\frac{1}{4} + \frac{2}{5}$

3. $\frac{1}{2} + \frac{1}{10}$ 6. $\frac{1}{2} + \frac{9}{10}$ 9. $\frac{1}{4} + \frac{3}{5}$

Exercise No. 296
Mental Multiplication
Multiply mentally the following.

1. 44×53
2. 54×54
3. 64×55
4. 74×56
5. 84×57
6. 94×58
7. 34×59

8. 44×61
9. 54×53
10. 64×54
11. 74×55
12. 84×56
13. 94×57
14. 34×58

15. 44×59
16. 59×61
17. 64×53
18. 74×54
19. 84×55
20. 94×56

Exercise No. 297
Subtraction of Fractions
Review the examples in Exercise No. 289 on page 108 and No. 293 on page 109. Also perform the following subtractions.

1. $\frac{5}{6} - \frac{1}{4}$
2. $1\frac{1}{6} - \frac{1}{4}$
3. $\frac{5}{6} - \frac{3}{4}$

4. $1\frac{1}{6} - \frac{3}{4}$
5. $1\frac{1}{3} - \frac{3}{4}$
6. $1\frac{2}{3} - \frac{3}{4}$

7. $\frac{5}{24} - \frac{1}{8}$
8. $\frac{13}{24} - \frac{1}{8}$
9. $\frac{17}{24} - \frac{1}{8}$

10. $1\frac{1}{24} - \frac{1}{8}$

Exercise No. 298
Mental Division
Divide mentally the following.

1. $3989 \div 754$
2. $4967 \div 865$
3. $5192 \div 976$
4. $1002 \div 181$
5. $1566 \div 292$
6. $4486 \div 696$
7. $4632 \div 747$

8. $5206 \div 851$
9. $6381 \div 962$
10. $1153 \div 173$
11. $982 \div 131$
12. $1829 \div 242$
13. $2706 \div 353$
14. $3433 \div 464$

15. $4089 \div 575$
16. $1200 \div 141$
17. $2141 \div 252$
18. $3084 \div 363$
19. $4152 \div 474$
20. $5101 \div 585$

Exercise No. 299
Addition of Fractions

Review the examples in Exercise No. 283 on page 104, No. 292 on page 108 and No. 295 on page 109. Also perform the following additions.

1. $\frac{1}{4} + \frac{1}{10}$
2. $\frac{1}{4} + \frac{3}{10}$
3. $\frac{1}{4} + \frac{7}{10}$

4. $\frac{1}{4} + \frac{9}{10}$
5. $\frac{3}{4} + \frac{1}{5}$
6. $\frac{3}{4} + \frac{2}{5}$

7. $\frac{3}{4} + \frac{3}{5}$
8. $\frac{3}{4} + \frac{4}{5}$
9. $\frac{3}{4} + \frac{1}{10}$

10. $\frac{3}{4} + \frac{3}{10}$

Exercise No. 300
Mental Multiplication

Multiply mentally the following.

1. 45×62
2. 55×63
3. 65×64
4. 75×65
5. 85×66
6. 95×67
7. 35×68

8. 45×69
9. 55×62
10. 65×63
11. 75×64
12. 85×65
13. 95×66
14. 35×67

15. 45×68
16. 55×69
17. 65×62
18. 75×63
19. 85×64
20. 95×65

Exercise No. 301
Subtraction of Fractions

Review the examples in Exercise No. 293 on page 109 and No. 297 on page 110. Also perform the following subtractions.

1. $\frac{11}{24} - \frac{3}{8}$
2. $\frac{19}{24} - \frac{3}{8}$
3. $\frac{23}{24} - \frac{3}{8}$

4. $1\frac{7}{24} - \frac{3}{8}$
5. $\frac{17}{24} - \frac{5}{8}$
6. $1\frac{11}{24} - \frac{5}{8}$

7. $1\frac{5}{24} - \frac{5}{8}$
8. $1\frac{13}{24} - \frac{5}{8}$
9. $\frac{23}{24} - \frac{7}{8}$

10. $1\frac{7}{24} - \frac{7}{8}$

Exercise No. 302
Mental Division

Divide mentally the following.

1. $1714 \div 284$
2. $2399 \div 395$

3. $2714 \div 446$
4. $3507 \div 557$

5. $4617 \div 661$
6. $5303 \div 686$

7. $5886 \div 797$ 12. $6588 \div 747$ 17. $2502 \div 263$
8. $6665 \div 838$ 13. $7189 \div 858$ 18. $3440 \div 374$
9. $7233 \div 941$ 14. $8238 \div 969$ 19. $4450 \div 485$
10. $1084 \div 152$ 15. $1385 \div 171$ 20. $5423 \div 596$
11. $5757 \div 696$ 16. $1493 \div 152$

Exercise No. 303
Addition of Fractions

Review the examples in Exercise No. 292 on page 108, No. 295 on page 109 and No. 299 on page 111. Also perform the following additions.

1. $\frac{3}{4} + \frac{7}{10}$ 4. $\frac{1}{8} + \frac{2}{5}$ 7. $\frac{1}{8} + \frac{1}{10}$ 10. $\frac{1}{8} + \frac{9}{10}$
2. $\frac{3}{4} + \frac{9}{10}$ 5. $\frac{1}{8} + \frac{3}{5}$ 8. $\frac{1}{8} + \frac{3}{10}$
3. $\frac{1}{8} + \frac{1}{5}$ 6. $\frac{1}{8} + \frac{4}{5}$ 9. $\frac{1}{8} + \frac{7}{10}$

Exercise No. 304
Mental Multiplication
Multiply mentally the following.

1. 46×71 8. 46×78 15. 46×77
2. 56×72 9. 56×71 16. 56×78
3. 66×73 10. 66×72 17. 66×71
4. 76×74 11. 76×73 18. 76×72
5. 86×75 12. 86×74 19. 86×73
6. 96×76 13. 96×75 20. 96×74
7. 36×77 14. 36×76

Exercise No. 305
Subtraction of Fractions

Review the examples in Exercise No. 297 on page 110 and No. 301 on page 111. Also perform the following subtractions.

1. $1\frac{11}{24} - \frac{7}{8}$ 4. $\frac{1}{2} - \frac{1}{5}$ 7. $\frac{1}{2} - \frac{2}{5}$ 10. $1\frac{3}{10} - \frac{2}{5}$
2. $1\frac{19}{24} - \frac{7}{8}$ 5. $\frac{9}{10} - \frac{1}{5}$ 8. $\frac{7}{10} - \frac{2}{5}$
3. $\frac{3}{10} - \frac{1}{5}$ 6. $1\frac{1}{10} - \frac{1}{5}$ 9. $1\frac{1}{10} - \frac{2}{5}$

Exercise No. 306
Mental Division
Divide mentally the following.

1. 5338 ÷ 772
2. 5393 ÷ 883
3. 6001 ÷ 994
4. 908 ÷ 145
5. 1576 ÷ 256
6. 1859 ÷ 263
7. 2736 ÷ 374

8. 3606 ÷ 485
9. 4518 ÷ 596
10. 4711 ÷ 637
11. 2284 ÷ 282
12. 3183 ÷ 393
13. 3956 ÷ 444
14. 4795 ÷ 555

15. 5954 ÷ 666
16. 5887 ÷ 647
17. 7123 ÷ 758
18. 8221 ÷ 869
19. 9257 ÷ 973
20. 1721 ÷ 184

Exercise No. 307
Addition of Fractions
Review the examples in Exercise No. 295 on page 109, No. 297 on page 110 and No. 303 on page 112. Also perform the following additions.

1. $\frac{3}{8} + \frac{1}{5}$
2. $\frac{3}{8} + \frac{2}{5}$
3. $\frac{3}{8} + \frac{3}{5}$

4. $\frac{3}{8} + \frac{4}{5}$
5. $\frac{3}{8} + \frac{1}{10}$
6. $\frac{3}{8} + \frac{3}{10}$

7. $\frac{3}{8} + \frac{7}{10}$
8. $\frac{3}{8} + \frac{9}{10}$
9. $\frac{5}{8} + \frac{1}{5}$

10. $\frac{5}{8} + \frac{2}{5}$

Exercise No. 308
Mental Multiplication
Perform mentally the following multiplications.

1. 47 × 79
2. 57 × 81
3. 67 × 82
4. 77 × 83
5. 87 × 84
6. 97 × 85
7. 37 × 86

8. 47 × 87
9. 57 × 79
10. 67 × 81
11. 77 × 82
12. 87 × 83
13. 97 × 84
14. 37 × 85

15. 47 × 86
16. 57 × 87
17. 67 × 79
18. 77 × 81
19. 87 × 82
20. 97 × 83

Exercise No. 309

Subtraction of Fractions

Review the examples in Exercise No. 301 on page 111 and No. 305 on page 112. Also perform the following subtractions.

1. $\frac{7}{10} - \frac{3}{5}$ 4. $1\frac{1}{2} - \frac{3}{5}$ 7. $1\frac{1}{2} - \frac{4}{5}$ 10. $\frac{9}{10} - \frac{1}{2}$

2. $\frac{9}{10} - \frac{3}{5}$ 5. $\frac{9}{10} - \frac{4}{5}$ 8. $1\frac{7}{10} - \frac{4}{5}$

3. $1\frac{3}{10} - \frac{3}{5}$ 6. $1\frac{1}{10} - \frac{4}{5}$ 9. $\frac{7}{10} - \frac{1}{2}$

Exercise No. 310

Mental Division

Divide mentally the following.

1. $5365 \div 748$ 8. $8304 \div 999$ 15. $6720 \div 679$

2. $6599 \div 851$ 9. $6075 \div 741$ 16. $7831 \div 784$

3. $7445 \div 962$ 10. $5241 \div 652$ 17. $8917 \div 895$

4. $1243 \div 173$ 11. $2682 \div 295$ 18. $9441 \div 946$

5. $2220 \div 284$ 12. $3411 \div 346$ 19. $1563 \div 157$

6. $6293 \div 777$ 13. $4471 \div 457$ 20. $2627 \div 268$

7. $7548 \div 888$ 14. $5667 \div 568$

Exercise No. 311

Addition of Fractions

Review the examples in Exercise No. 297 on page 110, No. 303 on page 112 and No. 307 on page 113. Also add the following.

1. $\frac{5}{8} + \frac{3}{5}$ 4. $\frac{5}{8} + \frac{3}{10}$ 7. $\frac{7}{8} + \frac{1}{5}$ 10. $\frac{7}{8} + \frac{4}{5}$

2. $\frac{5}{8} + \frac{4}{5}$ 5. $\frac{5}{8} + \frac{7}{10}$ 8. $\frac{7}{8} + \frac{2}{5}$

3. $\frac{5}{8} + \frac{1}{10}$ 6. $\frac{5}{8} + \frac{9}{10}$ 9. $\frac{7}{8} + \frac{3}{5}$

Exercise No. 312

Mental Multiplication

Multiply mentally the following.

1. 48 × 88	**8.** 48 × 96	**15.** 48 × 95
2. 58 × 89	**9.** 58 × 88	**16.** 58 × 96
3. 68 × 91	**10.** 68 × 89	**17.** 68 × 88
4. 78 × 92	**11.** 78 × 91	**18.** 78 × 89
5. 88 × 93	**12.** 88 × 92	**19.** 88 × 91
6. 98 × 94	**13.** 98 × 93	**20.** 98 × 92
7. 38 × 95	**14.** 38 × 94	

Exercise No. 313
Subtraction of Fractions

Review the examples in Exercise No. 305 on page 112 and No. 309 on page 114. Also perform the following subtractions.

1. $1\frac{1}{10} - \frac{1}{2}$ **4.** $\frac{4}{5} - \frac{1}{2}$ **7.** $\frac{9}{20} - \frac{1}{4}$ **10.** $1\frac{1}{20} - \frac{1}{4}$

2. $1\frac{3}{10} - \frac{1}{2}$ **5.** $1\frac{1}{5} - \frac{1}{2}$ **8.** $\frac{13}{20} - \frac{1}{4}$

3. $\frac{3}{5} - \frac{1}{2}$ **6.** $1\frac{2}{5} - \frac{1}{2}$ **9.** $\frac{17}{20} - \frac{1}{4}$

Exercise No. 314
Addition of Fractions

Review the examples in Exercise No. 303 on page 112, No. 307 on page 113 and No. 311 on page 114. Also perform the following additions.

1. $\frac{7}{8} + \frac{1}{10}$ **4.** $\frac{7}{8} + \frac{9}{10}$ **7.** $\frac{1}{3} + \frac{3}{5}$ **10.** $\frac{1}{3} + \frac{3}{10}$

2. $\frac{7}{8} + \frac{3}{10}$ **5.** $\frac{1}{3} + \frac{1}{5}$ **8.** $\frac{1}{3} + \frac{4}{5}$

3. $\frac{7}{8} + \frac{7}{10}$ **6.** $\frac{1}{3} + \frac{2}{5}$ **9.** $\frac{1}{3} + \frac{1}{10}$

Exercise No. 315
Mental Multiplication

Multiply the following mentally.

1. 49 × 95	**8.** 49 × 97	**15.** 49 × 99
2. 59 × 96	**9.** 59 × 98	**16.** 59 × 95
3. 69 × 97	**10.** 69 × 99	**17.** 69 × 96
4. 79 × 98	**11.** 79 × 95	**18.** 79 × 97
5. 89 × 99	**12.** 89 × 96	**19.** 89 × 98
6. 99 × 95	**13.** 99 × 97	**20.** 99 × 99
7. 39 × 96	**14.** 39 × 98	

Exercise No. 316
Subtraction of Fractions

Review the examples in Exercise No. 309 on page 114 and No. 313 on page 115. Also perform the following subtractions.

1. $\frac{7}{20} - \frac{1}{4}$
2. $\frac{11}{20} - \frac{1}{4}$
3. $\frac{19}{20} - \frac{1}{4}$
4. $1\frac{3}{20} - \frac{1}{4}$
5. $\frac{19}{20} - \frac{3}{4}$
6. $1\frac{3}{20} - \frac{3}{4}$
7. $1\frac{7}{20} - \frac{3}{4}$
8. $1\frac{11}{20} - \frac{3}{4}$
9. $\frac{17}{20} - \frac{3}{4}$
10. $1\frac{1}{20} - \frac{3}{4}$

Exercise No. 317
Addition of Fractions

Review the examples in Exercise No. 307 on page 113, No. 311 on page 114 and No. 314 on page 115. Also perform the following additions.

1. $\frac{1}{3} + \frac{7}{10}$
2. $\frac{1}{3} + \frac{9}{10}$
3. $\frac{2}{3} + \frac{1}{5}$
4. $\frac{2}{3} + \frac{2}{5}$
5. $\frac{2}{3} + \frac{3}{5}$
6. $\frac{2}{3} + \frac{4}{5}$
7. $\frac{2}{3} + \frac{1}{10}$
8. $\frac{2}{3} + \frac{3}{10}$
9. $\frac{2}{3} + \frac{7}{10}$
10. $\frac{2}{3} + \frac{9}{10}$

Exercise No. 318
Subtraction of Fractions

Review the examples in Exercise No. 313 on page 115 and No. 316 on this page. Also perform the following subtractions.

1. $1\frac{9}{20} - \frac{3}{4}$
2. $1\frac{13}{20} - \frac{3}{4}$
3. $\frac{13}{40} - \frac{1}{8}$
4. $\frac{21}{40} - \frac{1}{8}$
5. $\frac{29}{40} - \frac{1}{8}$
6. $\frac{37}{40} - \frac{1}{8}$
7. $\frac{9}{40} - \frac{1}{8}$
8. $\frac{17}{40} - \frac{1}{8}$
9. $\frac{33}{40} - \frac{1}{8}$
10. $1\frac{1}{40} - \frac{1}{8}$

Exercise No. 319
Mental Division

Divide the following mentally.

1. $1066 \div 26$
2. $1377 \div 27$
3. $1708 \div 28$
4. $2059 \div 29$
5. $2511 \div 31$
6. $2912 \div 32$

7. 1023 ÷ 33	**12.** 2349 ÷ 29	**17.** 1586 ÷ 26
8. 1394 ÷ 34	**13.** 2821 ÷ 31	**18.** 1917 ÷ 27
9. 1326 ÷ 26	**14.** 992 ÷ 32	**19.** 2268 ÷ 28
10. 1647 ÷ 27	**15.** 1353 ÷ 33	**20.** 2639 ÷ 29
11. 1988 ÷ 28	**16.** 1734 ÷ 34	

Exercise No. 320
Addition of Fractions

Review the examples in Exercise No. 311 on page 114, No. 314 on page 115 and No. 315 on page 115. Also perform the following additions.

1. $\frac{1}{6} + \frac{1}{5}$ **4.** $\frac{1}{6} + \frac{4}{5}$ **7.** $\frac{1}{6} + \frac{7}{10}$ **10.** $\frac{5}{6} + \frac{2}{5}$

2. $\frac{1}{6} + \frac{2}{5}$ **5.** $\frac{1}{6} + \frac{1}{10}$ **8.** $\frac{1}{6} + \frac{9}{10}$

3. $\frac{1}{6} + \frac{3}{5}$ **6.** $\frac{1}{6} + \frac{3}{10}$ **9.** $\frac{5}{6} + \frac{1}{5}$

Exercise No. 321
Subtraction of Fractions

Review the examples in Exercise No. 314 on page 115, No. 316 on page 116 and No. 320 above. Also perform the following subtractions.

1. $\frac{23}{40} - \frac{3}{8}$ **4.** $1\frac{7}{40} - \frac{3}{8}$ **7.** $1\frac{3}{40} - \frac{3}{8}$ **10.** $1\frac{1}{40} - \frac{5}{8}$

2. $\frac{31}{40} - \frac{3}{8}$ **5.** $\frac{19}{40} - \frac{3}{8}$ **8.** $1\frac{11}{40} - \frac{3}{8}$

3. $\frac{39}{40} - \frac{3}{8}$ **6.** $\frac{27}{40} - \frac{3}{8}$ **9.** $\frac{33}{40} - \frac{5}{8}$

Exercise No. 322
Mental Division

Divide the following mentally.

1. 1470 ÷ 35	**8.** 1806 ÷ 43	**15.** 1764 ÷ 42
2. 1872 ÷ 36	**9.** 1820 ÷ 35	**16.** 2236 ÷ 43
3. 2294 ÷ 37	**10.** 2232 ÷ 36	**17.** 2108 ÷ 34
4. 2736 ÷ 38	**11.** 2664 ÷ 37	**18.** 2520 ÷ 35
5. 3198 ÷ 39	**12.** 3116 ÷ 38	**19.** 2952 ÷ 36
6. 3772 ÷ 41	**13.** 3588 ÷ 39	**20.** 3404 ÷ 37
7. 1344 ÷ 42	**14.** 1312 ÷ 41	

Exercise No. 323
Addition of Fractions

Review the examples in Exercise No. 314 on page 115, No. 317 on page 116 and No. 320 on page 117. Also perform the following additions.

1. $\frac{5}{6} + \frac{3}{5}$ 3. $\frac{5}{6} + \frac{1}{10}$ 5. $\frac{5}{6} + \frac{7}{10}$
2. $\frac{5}{6} + \frac{4}{5}$ 4. $\frac{5}{6} + \frac{3}{10}$ 6. $\frac{5}{6} + \frac{9}{10}$

Exercise No. 324
Subtraction of Fractions

Review the examples in Exercise No. 318 on page 116 and No. 321 on page 117. Also perform the following subtractions.

1. $1\frac{9}{40} - \frac{5}{8}$ 4. $\frac{37}{40} - \frac{5}{8}$ 7. $1\frac{3}{40} - \frac{7}{8}$ 10. $1\frac{27}{40} - \frac{7}{8}$
2. $1\frac{17}{40} - \frac{5}{8}$ 5. $1\frac{11}{40} - \frac{5}{8}$ 8. $1\frac{11}{40} - \frac{7}{8}$
3. $\frac{29}{40} - \frac{5}{8}$ 6. $1\frac{21}{40} - \frac{5}{8}$ 9. $1\frac{19}{40} - \frac{7}{8}$

Exercise No. 325
Mental Division

Divide the following mentally.

1. $1892 \div 44$ 8. $2236 \div 52$ 15. $2193 \div 51$
2. $2385 \div 45$ 9. $2332 \div 44$ 16. $2756 \div 52$
3. $2898 \div 46$ 10. $2835 \div 45$ 17. $2772 \div 44$
4. $3431 \div 47$ 11. $3358 \div 46$ 18. $3285 \div 45$
5. $3984 \div 48$ 12. $3901 \div 47$ 19. $3818 \div 46$
6. $4557 \div 49$ 13. $4464 \div 48$ 20. $4371 \div 47$
7. $1683 \div 51$ 14. $1617 \div 49$

Exercise No. 326
Addition of Fractions

Review the examples in Exercise No. 317 on page 116, No. 320 on page 117 and No. 323 on this page.

Exercise No. 327
Subtraction of Fractions

Review the examples in Exercise No. 321 on page 117 and No. 324 on page 118. Also perform the following subtractions.

1. $\frac{39}{40} - \frac{7}{8}$ **4.** $1\frac{31}{40} - \frac{7}{8}$ **7.** $\frac{14}{15} - \frac{1}{3}$ **10.** $\frac{19}{30} - \frac{1}{3}$

2. $1\frac{7}{40} - \frac{7}{8}$ **5.** $\frac{8}{15} - \frac{1}{3}$ **8.** $1\frac{2}{15} - \frac{1}{3}$

3. $1\frac{23}{40} - \frac{7}{8}$ **6.** $\frac{11}{15} - \frac{1}{3}$ **9.** $\frac{13}{30} - \frac{1}{3}$

Exercise No. 328
Mental Division

Divide the following mentally.

1. $2332 \div 53$ **8.** $2684 \div 61$ **15.** $2596 \div 59$

2. $2916 \div 54$ **9.** $2862 \div 53$ **16.** $3294 \div 61$

3. $3520 \div 55$ **10.** $3456 \div 54$ **17.** $3392 \div 53$

4. $4144 \div 56$ **11.** $4070 \div 55$ **18.** $3996 \div 54$

5. $4788 \div 57$ **12.** $4704 \div 56$ **19.** $4620 \div 55$

6. $5452 \div 58$ **13.** $5358 \div 57$ **20.** $5264 \div 56$

7. $2006 \div 59$ **14.** $1972 \div 58$

Exercise No. 329
Addition of Fractions

Review the examples in Exercise No. 320 on page 117 and 323 on page 118.

Exercise No. 330
Subtraction of Fractions

Review the examples in Exercise No. 321 on page 117 and No. 324 on page 118. Also perform the following subtractions.

1. $1\frac{1}{30} - \frac{1}{3}$ **4.** $1\frac{1}{15} - \frac{2}{3}$ **7.** $\frac{23}{30} - \frac{2}{3}$ **10.** $1\frac{17}{30} - \frac{2}{3}$

2. $1\frac{7}{30} - \frac{1}{3}$ **5.** $1\frac{4}{15} - \frac{2}{3}$ **8.** $\frac{29}{30} - \frac{2}{3}$

3. $\frac{13}{15} - \frac{2}{3}$ **6.** $1\frac{7}{15} - \frac{2}{3}$ **9.** $1\frac{11}{30} - \frac{2}{3}$

Exercise No. 331
Mental Division
Divide the following mentally.

1. 2790 ÷ 62
2. 3465 ÷ 63
3. 4160 ÷ 64
4. 4875 ÷ 65
5. 5610 ÷ 66
6. 6365 ÷ 67
7. 2380 ÷ 68

8. 3105 ÷ 69
9. 3410 ÷ 62
10. 4095 ÷ 63
11. 4800 ÷ 64
12. 5525 ÷ 65
13. 6270 ÷ 66
14. 2345 ÷ 67

15. 3060 ÷ 68
16. 3795 ÷ 69
17. 4030 ÷ 62
18. 4725 ÷ 63
19. 5440 ÷ 64
20. 6175 ÷ 65

Exercise No. 332
Mental Division
Divide the following mentally.

1. 3266 ÷ 71
2. 4032 ÷ 72
3. 4818 ÷ 73
4. 5624 ÷ 74
5. 6450 ÷ 75
6. 7296 ÷ 76
7. 2772 ÷ 77

8. 3588 ÷ 78
9. 3976 ÷ 71
10. 4752 ÷ 72
11. 5548 ÷ 73
12. 6364 ÷ 74
13. 7200 ÷ 75
14. 2736 ÷ 76

15. 3542 ÷ 77
16. 4368 ÷ 78
17. 4686 ÷ 71
18. 5472 ÷ 72
19. 6278 ÷ 73
20. 7104 ÷ 74

Exercise No. 333
Subtraction of Fractions
Review the examples in Exercise No. 324 on page 118 and No. 330 on page 119. Also perform the following subtractions.

1. $\frac{11}{30} - \frac{1}{6}$
2. $\frac{17}{30} - \frac{1}{6}$
3. $\frac{23}{30} - \frac{1}{6}$

4. $\frac{29}{30} - \frac{1}{6}$
5. $\frac{4}{15} - \frac{1}{6}$
6. $\frac{7}{15} - \frac{1}{6}$

7. $\frac{13}{15} - \frac{1}{6}$
8. $1\frac{1}{15} - \frac{1}{6}$
9. $1\frac{1}{30} - \frac{5}{6}$

10. $1\frac{7}{30} - \frac{5}{6}$

Exercise No. 334
Mental Division
Divide the following mentally.

1. 3713 ÷ 79
2. 4617 ÷ 81
3. 5494 ÷ 82

4. 6391 ÷ 83
5. 7308 ÷ 84
6. 8245 ÷ 85

7. 3182 ÷ 86
8. 4089 ÷ 87
9. 4503 ÷ 79

10. 5427 ÷ 81   **14.** 3145 ÷ 85   **18.** 6237 ÷ 81
11. 6314 ÷ 82   **15.** 4042 ÷ 86   **19.** 7134 ÷ 82
12. 7221 ÷ 83   **16.** 4959 ÷ 87   **20.** 8051 ÷ 83
13. 8148 ÷ 84   **17.** 5293 ÷ 79

Exercise No. 335
Subtraction of Fractions

Review the examples in Exercise No. 330 on page 119 and No. 333 on page 120. Also perform the following subtractions.

1. $1\frac{13}{30} - \frac{5}{6}$ **3.** $\frac{14}{15} - \frac{5}{6}$ **5.** $1\frac{8}{15} - \frac{5}{6}$
2. $1\frac{19}{30} - \frac{5}{6}$ **4.** $1\frac{2}{15} - \frac{5}{6}$ **6.** $1\frac{11}{15} - \frac{5}{6}$

Exercise No. 336
Mental Division

Divide the following mentally.

1. 4224 ÷ 88   **8.** 4608 ÷ 96   **15.** 4560 ÷ 95
2. 5162 ÷ 89   **9.** 5104 ÷ 88   **16.** 5568 ÷ 96
3. 6188 ÷ 91   **10.** 6052 ÷ 89   **17.** 5984 ÷ 88
4. 7176 ÷ 92   **11.** 7098 ÷ 91   **18.** 6942 ÷ 89
5. 8184 ÷ 93   **12.** 8096 ÷ 92   **19.** 8008 ÷ 91
6. 9212 ÷ 94   **13.** 9114 ÷ 93   **20.** 9016 ÷ 92
7. 3610 ÷ 95   **14.** 3572 ÷ 94

Exercise No. 337
Mental Division

Divide the following mentally.

1. 4655 ÷ 95   **8.** 4753 ÷ 97   **15.** 4851 ÷ 99
2. 5664 ÷ 96   **9.** 5782 ÷ 98   **16.** 5605 ÷ 95
3. 6693 ÷ 97   **10.** 6831 ÷ 99   **17.** 6624 ÷ 96
4. 7742 ÷ 98   **11.** 7505 ÷ 95   **18.** 7663 ÷ 97
5. 8811 ÷ 99   **12.** 8544 ÷ 96   **19.** 8722 ÷ 98
6. 9405 ÷ 95   **13.** 9603 ÷ 97   **20.** 9801 ÷ 99
7. 3744 ÷ 96   **14.** 3822 ÷ 98

DECIMALS IN GENERAL

For the purposes of this book our interest in decimals centers in the equivalence of value between certain decimals and common fractions. Decimal parts of a number that may be represented as simple fractions of that number are known as *aliquot parts* of it. Thus, $12\frac{1}{2}$, 25 and $33\frac{1}{3}$ are aliquot parts of 100, being respectively equal to $\frac{1}{8}$, $\frac{1}{4}$ and $\frac{1}{3}$ of 100.

A knowledge of aliquot parts simplifies many arithmetical calculations. Thus if it be required to multiply 7928 by 25, the simplest way is to annex two 0's to 7928, making it 792800, and then divide by 4, since 25 is $\frac{1}{4}$ of 100. The answer, which may easily be figured mentally, comes to 198200.

Again, if we wanted to know the cost of 25 gross of penholders at $66\frac{2}{3}$¢ per dozen, we would figure that 1 gross costs $\$\frac{2}{3} \times 12$, or \$8, and that 25 gross therefore cost \$200.

Everybody with any degree of arithmetical training or experience is familiar with the equivalent decimal values for halves, quarters, eighths, thirds, sixths, fifths, tenths, twentieths, twenty-fifths and fiftieths. It is not difficult to extend the list of memorized values so as to include sixteenths and twelfths, and with this knowledge to make rapid calculations of values in thirty-seconds and twenty-fourths.

The succeeding exercises in decimals are designed toward this end. The student is drilled in representing the values of various fractions as decimals of an increasingly higher number of places. No tables are given because values are more quickly learned by repeated calculation than by any effort at mere memorization.

Exercise No. 338
Two-Place Decimal Values

Express the following fractions as decimals of two places. Use fractional terminations where necessary. Thus, $\frac{1}{3}$ expressed as a two-place decimal becomes $.33\frac{1}{3}$.

1. $\frac{1}{8}$	**4.** $\frac{7}{8}$	**7.** $\frac{1}{6}$	**10.** $\frac{2}{5}$
2. $\frac{3}{8}$	**5.** $\frac{1}{3}$	**8.** $\frac{5}{8}$	**11.** $\frac{3}{5}$
3. $\frac{5}{8}$	**6.** $\frac{2}{3}$	**9.** $\frac{1}{5}$	**12.** $\frac{4}{5}$

Repeat this exercise three times.

Exercise No. 339
Multiplying Three Figures by Two

Multiply mentally the following.

No new principles are involved in multiplications of this type. The student is simply asked to apply the methods which he has already learned to larger numbers.

1. 111×26	**4.** 442×29	**7.** 721×33	**10.** 152×27
2. 222×27	**5.** 551×31	**8.** 832×34	
3. 331×28	**6.** 612×32	**9.** 941×26	

Exercise No. 340
Two-Place Decimal Values

Review the examples in Exercise No. 338 above.

Express the following as decimals of two places.

1. $\frac{1}{16}$	**5.** $\frac{9}{16}$	**9.** $\frac{1}{12}$	**13.** $\frac{1}{32}$
2. $\frac{3}{16}$	**6.** $\frac{11}{16}$	**10.** $\frac{5}{12}$	**14.** $\frac{1}{24}$
3. $\frac{5}{16}$	**7.** $\frac{13}{16}$	**11.** $\frac{7}{12}$	
4. $\frac{7}{16}$	**8.** $\frac{15}{16}$	**12.** $\frac{11}{12}$	

Repeat this exercise three times.

Exercise No. 341
Multiplying Three Figures by Two
Multiply mentally the following.

1. 121 × 35	4. 451 × 38	7. 731 × 42	10. 161 × 36
2. 232 × 36	5. 562 × 39	8. 842 × 43	
3. 343 × 37	6. 623 × 41	9. 953 × 35	

SHORT CUTS

There are a number of devices for shortening the work of calculation in specific cases, though most of the methods usually included under this head have only a limited practical value because they are applicable only in highly special cases. A few methods, like horizontal addition and combined addition and subtraction have first-class utility. A variety of short cuts of varying degrees of value are given in the following pages without any attempt to classify them. The student should become familiar with all of them because there is always benefit in viewing numbers from as many angles as possible.

Exercise No. 342
Horizontal Addition

The term *horizontal addition* is applied to the adding of numbers that are not arranged in column form. There is often an unnecessary waste of time in arranging numbers in the form of columns. This is particularly true when the numbers to be added are on bills, invoices, etc. Values on such papers may be totalled by writing down each partial sum as it is arrived at, and then making a final addition.

Consider the first of the following examples. The sum of the units is 37, the sum of the tens is 45, etc. The sums of the various orders are successively set down in the form shown below, and then added.

$$37$$
$$45$$
$$14$$
$$\underline{16}$$
$$17887$$

The process might of course be shortened somewhat by adding two orders at a time.

Add the following.

1. $32 + $183 + $54 + $3486 + $569 + $9375 + $85 + $4103
2. $875 + $284 + $37 + $5200 + $398 + $62 + $74 + $2168 + $720
3. 763 + 827 + 49 + 5283 + 768 + 2175
4. 1536 + 8973 + 5178 + 926 + 8259 + 36 + 867
5. 9365 + 8375 + 1473 + 826 + 4123 + 15378
6. 986 + 325 + 7261 + 5820 + 569 + 8371
7. 6275 + 5183 + 985 + 3267 + 75 + 1528
8. 1738 + 9168 + 8273 + 5298 + 9 + 6832 + 65
9. $783.52 + $41.27 + $837.45 + $9681.73 + $48.26 + $912.78 + $91.75 + $683.12 + $41.83 + $591.87 + $291.83 + $758.32 + $58.67
10. 46235 + 8976 + 5807 + 98397 + 68325 + 892 + 5140 + 6839 + 326 + 2125

Exercise No. 343

Multiplying Three Figures by Two

Multiply mentally the following.

1. 131×44	4. 464×47	7. 743×51	10. 172×45
2. 242×45	5. 571×48	8. 854×52
3. 353×46	6. 632×49	9. 961×44

Exercise No. 344

Four-Place Decimal Values

Review the examples in Exercises No. 338 and 340 on page 123.

Express the fractions listed in Exercise No. 340 as decimals of four places. This is done by simply writing the value as parts of 100 of the terminal fractions of the proper two-place decimals. Thus, $\frac{1}{16}$, which is $.06\frac{1}{4}$ as a two-place decimal, becomes .0625 as a decimal of four places. Again, $\frac{1}{12}$ is $.08\frac{1}{3}$ or $.0833\frac{1}{3}$.

Exercise No. 345
Multiplying Three Figures by Two
Multiply mentally the following.

1. 141 × 53   4. 474 × 56   7. 752 × 59   10. 185 × 54
2. 252 × 54   5. 585 × 57   8. 863 × 61
3. 363 × 55   6. 641 × 58   9. 974 × 53

Exercise No. 346
Combined Addition and Subtraction

It sometimes becomes necessary to subtract the sum of several numbers from a single number. If the numbers to be added are arranged in column form, this may be done at what amounts to one operation by a very simple process.

The numbers may be arranged either as a sum with a missing addend, as in the examples given for practice, or else with the minuend written at the top with underscoring and the difference written at the bottom, as in the examples shown for illustration.

The so-called carry method of subtraction is used. The sum of each successive column is subtracted from the corresponding figure of the minuend plus as many tens as may be necessary to make the subtraction possible. The number of tens thus used is then added to the next column.

To illustrate: from 122808 take the sum of 35635, and 68921.

122808

35635
68921

18252

The sum of 5 and 1 is subtracted from 8; write 2 and carry 0. Subtract 5 from 10; write 5 and carry 1 because 1 ten was used to make the subtraction possible. With

1 to carry, the next column adds to 16; subtract this from 18 and again carry 1. The next column adds to 14; subtract this from 22 and carry 2 because 2 tens were needed to make the subtraction possible in this case. Carrying 2 and subtracting from 12 gives the final necessary figure, 1.

The method of carrying may be made still more clear by taking an example that involves larger numbers; from 3744 subtract the sum of 366, 466, 566, 666, 766, 266 and 466.

$$
\begin{array}{r}
3744 \\
\hline
366 \\
466 \\
566 \\
666 \\
766 \\
266 \\
466 \\
\hline
182
\end{array}
$$

The sum of the first column, 42, is subtracted from 44 because 44 is the next higher number ending in 4 from which a subtraction can be made; 4 is carried. The sum of the second column, 46, is subtracted from 54 because 54 is the next higher number ending in 4 from which a subtraction can be made; 5 is carried. The sum of the hundreds' column subtracted from 39 leaves 1.

In the following examples fill in in each case the missing number that will make all the numbers add to the total shown.

1.	2.	3.	4.
$24.96	6016	$29.44	6144
6.24	376	7.36	384
1.56	141	1.84	24576
12.48	188	3.68	3072
.98	1504	58.88	145
3.12	752	1.38	49152
(?)	(?)	(?)	(?)
$149.18	105233	$220.34	181777

5.	864	6.	$168.86	7.	$475.17	8.	$286.09
	108		10.56		46.82		5304.62
	81		1.32		120.08		20463.20
	5296		.96		2461.50		607.05
	3456		2.64		500.07		6315.46
	432		84.48		1208.92		73.90
	(?)		(?)		(?)		(?)
	11965		$944.66		$12933.16		$63452.87

Exercise No. 347

Multiplying Three Figures by Two

Multiply mentally the following.

1. 151 × 62   **4.** 484 × 65   **7.** 761 × 68   **10.** 194 × 63
2. 262 × 63   **5.** 595 × 66   **8.** 872 × 69
3. 373 × 64   **6.** 656 × 67   **9.** 983 × 62

Exercise No. 348

Five-Place Decimal Values

Review the examples in Exercises No. 338 and 340 on page 123 and No. 344 on page 126.

Express the following fractions as decimals of five places.

To find values in thirty-seconds, add .0312½ to the next lower value in sixteenths, etc. The calculation is clearer in the mind if both sixteenths and thirty-seconds are first thought of as decimals of four places. Changing the four-place answer to five places is the work of an instant.

To find values in twenty-fourths, add .0416⅔ to the next lower value in twelfths, etc. In writing answers, drop final ⅓, and raise final ⅔ to make the last figure a 7.

1. $\frac{1}{32}$ **4.** $\frac{7}{32}$ **7.** $\frac{13}{32}$ **10.** $\frac{19}{32}$ **13.** $\frac{25}{32}$
2. $\frac{3}{32}$ **5.** $\frac{9}{32}$ **8.** $\frac{15}{32}$ **11.** $\frac{21}{32}$ **14.** $\frac{27}{32}$
3. $\frac{5}{32}$ **6.** $\frac{11}{32}$ **9.** $\frac{17}{32}$ **12.** $\frac{23}{32}$ **15.** $\frac{29}{32}$

16. $\frac{31}{32}$ **18.** $\frac{5}{24}$ **20.** $\frac{11}{24}$ **22.** $\frac{17}{24}$ **24.** $\frac{23}{24}$

17. $\frac{1}{24}$ **19.** $\frac{7}{24}$ **21.** $\frac{13}{24}$ **23.** $\frac{19}{24}$

Exercise No. 349

Multiplying Three Figures by Two

Multiply mentally the following.

1. 141 × 71   **4.** 474 × 74   **7.** 747 × 77   **10.** 173 × 72
2. 252 × 72   **5.** 585 × 75   **8.** 851 × 78
3. 363 × 73   **6.** 696 × 76   **9.** 962 × 71

Exercise No. 350

Multiplying by a Near Number

It sometimes happens that a multiplier is a little more or a little less than 100, 1000, 10000, etc. In cases of this kind it is quickest to multiply by the round number and then add or subtract the necessary difference. For example, multiply $385.20 by 998. We multiply the dollar value by 1000 and subtract from this product twice $385.20, thus:

$$\$385200$$
$$\underline{\quad770.40\quad}$$

$$\$384429.60$$

Multiply the following. The student should be able to do most of these mentally.

1. $425 × 999   **4.** $258.30 × 104   **7.** $989 × 992
2. $865 × 98   **5.** $827.58 × 1003   **8.** $99 × 97
3. $735.25 × 998   **6.** $516 × 1.02   **9.** $1005 × 1002

Exercise No. 351
Multiplying Three Figures by Two
Multiply mentally the following.

1. 131 × 79   4. 464 × 83   7. 797 × 86   10. 152 × 81
2. 242 × 81   5. 575 × 84   8. 838 × 87
3. 353 × 82   6. 686 × 85   9. 941 × 79

Exercise No. 352
Review of Decimals

Review the examples in Exercise No. 340 on page 123, No. 344 on page 126 and No. 348 on page 129.

Exercise No. 353
Multiplying Three Figures by Two

Multiply mentally the following.

1. 141 × 88   4. 474 × 92   7. 747 × 95   10. 171 × 89
2. 252 × 89   5. 585 × 93   8. 858 × 96
3. 363 × 91   6. 696 × 94   9. 969 × 88

Exercise No. 354
Aliquot Parts in Multiplication

Reference has already been made to the fact that multiplication may be simplified by considering one of the factors as an aliquot part of some number ending in two or more 0's. Thus, 628 × 25 would be solved by multiplying 628 by 100 and dividing by 4; the answer comes to 15700. Again, multiplying 56 × 75 would be done most quickly by taking $\frac{3}{4}$ of 56 and then multiplying by 100.

Perform the following multiplications by the method of aliquot parts.

1. $35 × 15   6. $36 × 25   11. $35 × 18
2. $42 × 18   7. $52 × 250   12. $28 × 450
3. $24 × 16   8. $42 × 350   13. $36 × 33$\frac{1}{3}$
4. $18 × 45   9. $150 × 48   14. $72 × 16$\frac{2}{3}$
5. $72 × 75   10. $64 × 25   15. $96 × 12$\frac{1}{2}$

Exercise No. 355
Multiplying Three Figures by Two
Multiply mentally the following. Do not use short cuts.

1. 152×95 4. 485×98 7. 758×96 10. 194×99
2. 263×96 5. 596×99 8. 869×97
3. 374×97 6. 647×95 9. 973×98

Exercise No. 356
Review of Decimals
Review the examples in Exercise No. 344 on page 126 and No. 348 on page 129.

Exercise No. 357
Multiplying Three Figures by Three
Multiply mentally the following. Add together the first two partial products before determining the third.

1. 111×101 5. 551×141 9. 941×181
2. 222×111 6. 612×151 10. 152×191
3. 331×121 7. 721×161
4. 442×131 8. 832×171

Exercise No. 358
Simplifying the Multiplier
Sometimes a multiplier is of such a nature that one part of it may be taken as an exact multiple of another. In such cases an operation is eliminated by making a single multiplication of the first-found partial product instead of two multiplications of the original multiplicand. In the example at the left above, the 18 in the multiplier is equal to 3 times the 6. We therefore multiply the first partial product by 3 instead of multiplying the original multiplicand by 18. In the example at the right, 56 being equal

to 8 times 7, we multiply first by 8, placing the result in
the proper position, and then multiply this partial product
by 7.

$$
\begin{array}{r} 2574 \\ 186 \\ \hline 15444 \\ 46332 \\ \hline 478764 \end{array}
\qquad
\begin{array}{r} 5462 \\ 856 \\ \hline 43696 \\ 305872 \\ \hline 4675472 \end{array}
$$

Multiply the following by this method.

1. $385.85 × 642
2. $742.50 × 328
3. $82615 × 729
4. $4265.25 × 255

5. $9541.12 × 546
6. $172.48 × 763
7. $2153.28 × 18624
8. $530.75 × 16412

Exercise No. 359
Multiplying Three Figures by Three
Multiply mentally the following.

1. 121 × 202
2. 232 × 212
3. 343 × 222
4. 451 × 232

5. 562 × 242
6. 623 × 252
7. 731 × 262
8. 842 × 272

9. 953 × 282
10. 161 × 292

Exercise No. 360
Review of Decimals
Review the examples in Exercise No. 348 on page 129.

Exercise No. 361
Multiplying Three Figures by Three
Multiply mentally the following.

1. 131 × 303
2. 242 × 313
3. 353 × 323
4. 464 × 333

5. 571 × 343
6. 632 × 353
7. 743 × 363
8. 854 × 373

9. 961 × 383
10. 172 × 393

Exercise No. 362
Multiplication by Factoring

When a multiplier can be taken as the product of two factors, it may be quicker to make separate multiplications by each of these factors than to proceed in the ordinary manner. Take the example 632 × 156. In the illustrations below, the one at the left shows the ordinary method. At the right the multiplier is split up into the factors 13 and 12; the multiplicand is multiplied by 13 and the result is then multiplied by 12.

$$
\begin{array}{r}
632 \\
156 \\
\hline
3792 \\
3160 \\
632 \\
\hline
98592
\end{array}
\qquad
\begin{array}{r}
632 \\
13 \\
\hline
8216 \\
12 \\
\hline
98592
\end{array}
$$

Multiply the following by this method.

1. 759 × 182   **4.** 656 × 285   **7.** 542 × 221
2. 684 × 169   **5.** 309 × 289   **8.** 327 × 224
3. 327 × 228   **6.** 728 × 324   **9.** 986 × 196

Exercise No. 363
Multiplying Three Figures by Three
Multiply mentally the following.

1. 141 × 404   **5.** 585 × 444   **9.** 974 × 484
2. 252 × 414   **6.** 641 × 454   **10.** 185 × 494
3. 363 × 424   **7.** 752 × 464
4. 474 × 434   **8.** 863 × 474

Exercise No. 364
Factors Between 11 and 19

A quick way to calculate the product of two numbers between 11 and 19 is to add the units of one number to the whole of the other, annex 0 and add the product of the units of both numbers. Thus, to multiply 16 × 18:

16 and 8 are 24; call this 240 and add 48, making 288.
The same result would be reached by adding 6 to 18.
Multiply by this method:

1. 14 × 15	**4.** 15 × 16	**7.** 16 × 17
2. 18 × 19	**5.** 13 × 15	**8.** 14 × 16
3. 15 × 17	**6.** 13 × 19	**9.** 19 × 19

Exercise No. 365
Multiplying Three Figures by Three
Multiply mentally the following.

1. 151 × 505	**5.** 595 × 545	**9.** 983 × 585
2. 262 × 515	**6.** 656 × 555	**10.** 194 × 595
3. 373 × 525	**7.** 761 × 565	
4. 484 × 535	**8.** 872 × 575	

Exercise No. 366
Multiplying by 11
When the multiplicand consists of two figures the sum
of which is less than 10, the product is found by writing
the two figures of the multiplicand with their sum between
them. Thus, to multiply 62 by 11 we write 6 and 2 with
the sum of 6 and 2 between these figures, obtaining 682.

To multiply larger numbers by 11, apply the following
rule. Beginning at the right, write the units' figure of the
multiplicand, then successively the units plus the tens,
the tens plus the hundreds, the hundreds plus the thousands, etc., carrying wherever necessary, and ending with
the highest order of the multiplicand, or the highest order
plus the carrying figure. Thus, to multiply 4762 by 11:
write 2; add 2 and 6 and write 8; add 6 and 7, write 3 and
carry 1; add 7 and 4, increase it by the 1 carried, write 2
and carry 1; add this 1 to 4 and write 5. Answer, 52382.

Multiply the following by this method.

1. $5136 × 11   5. $41268.45 × 11
2. $72638 × 11   6. $3275.75 × 11
3. $514832 × 11   7. $48263.25 × 11
4. $37281.05 × 11   8. $94873.30 × 11

Exercise No. 367
Multiplying Three Figures by Three
Multiply mentally the following.

1. 141 × 606   5. 585 × 646   9. 962 × 686
2. 252 × 616   6. 696 × 656   10. 173 × 696
3. 363 × 626   7. 747 × 666
4. 474 × 636   8. 851 × 676

Exercise No. 368
Multiplying by 21, 31, 41, etc.

Setting down the product from right to left, write the units' figure of the multiplicand, then multiply each order of the multiplicand by the tens' figure of the multiplier, increasing the result in each case by the next higher order of the multiplicand and any necessary carrying figure.

Example, multiply 387 by 41; write 7; multiply 7 by 4, add the 8 of the multiplicand, making 36, write 6 and carry 3; multiply 8 by 4, add the 3 of the multiplicand and the carried 3, making 38, write 8 and carry 3; multiply 3 by 4 and add the carried 3 making 15, write 15. Answer, 15867.

Multiply by this method:

1. $2735.50 × 51   5. $7415.40 × 61
2. $1824.75 × 81   6. $8291.25 × 91
3. $5104.30 × 31   7. $2134.15 × 71
4. $6238.65 × 21   8. $5827.80 × 41

Exercise No. 369
Multiplying Three Figures by Three
Multiply mentally the following.

1. 131 × 707
2. 242 × 717
3. 353 × 727
4. 464 × 737

5. 575 × 747
6. 686 × 757
7. 797 × 767
8. 838 × 777

9. 941 × 787
10. 152 × 797

Exercise No. 370
Squares of Numbers

The square of a number is the number multiplied by itself. Squares may be determined quickly if the given number is considered to be the sum of two numbers. In algebra such a sum would ordinarily be taken as $a + b$ and its square would be $a^2 + 2 ab + b^2$. In regular arithmetical cases a becomes the tens of the number and b the units. Thus, 25 is 20 + 5, and 146 is 140 + 6. The algebraic formula for the square of the sum of two numbers is expressed as the square of the first plus twice the product of the first by the second plus the square of the second. Thus, 25 squared is 20 × 20 (400) plus 2 × 20 × 5 (200) plus 5 × 5 (25); the total is 625.

In computing squares by this principle you may immediately annex the square of the second to the square of the first, and then add twice the product of the first by the second. Thus in squaring 25 you would immediately say 425, and then add to this 2 × 20 × 5 (200), making 625. In squaring 146 you immediately say 19636 and add to this 2 × 140 × 6 (1680), making 21316. Always allow two places for the square of the second. Thus in squaring 61 the first partial product is 3601, to which 120 is added to make 3721.

In squaring numbers on paper the following method will be found rapid where large numbers are involved. Set the given number down twice as if for regular multiplication. Assuming that it is considered to consist of tens and units, multiply units by units, write units in the result and carry the tens. Add the two given tens together, multiply this sum by the given units, add the carried figure, write tens in the result and carry hundreds. Multiply tens by tens, add the carried figure and write the result.

67	134	1613
67	134	1613
4489	17956	2601769

In the first illustrative example at the left, $7 \times 7 = 49$, write 9 and carry 4; $6 + 6 = 12$, $12 \times 7 = 84$, $84 + 4 = 88$, write 8 and carry 8; $6 \times 6 = 36$, $36 + 8 = 44$.

In the second example, $4 \times 4 = 16$, write 6 and carry 1; $13 + 13 = 26$, $26 \times 4 = 104$, $104 + 1 = 105$, write 5 and carry 10; $13 \times 13 = 169$, $169 + 10 = 179$, write 179.

The third example is worked somewhat differently because here the parts of the number are considered to be 1600 and 13. $13 \times 13 = 169$, write 69 (two figures) and carry 1; $16 + 16 = 32$, $32 \times 13 = 416$, $416 + 1 = 417$, write 17 and carry 4; $16 \times 16 = 256$, $256 + 4 = 260$, write 260.

Find the squares of the following numbers. Do all the examples first by the first method, then by the second method.

1. 74	**4.** 64	**7.** 124	**10.** 197	**13.** 1314
2. 93	**5.** 38	**8.** 146	**11.** 1112	**14.** 1516
3. 82	**6.** 112	**9.** 168	**12.** 1213	**15.** 1719

Exercise No. 371
Multiplying Three Figures by Three
Multiply mentally the following.

1. 141×808	**5.** 585×848	**9.** 969×888
2. 252×818	**6.** 696×858	**10.** 171×898
3. 363×828	**7.** 747×868	
4. 474×838	**8.** 858×878	

Exercise No. 372
Multiplying When Units Are Alike

The following method is a variation of that explained in connection with the squaring of numbers.

```
    47              613
    67              913
  3149            559669
```

In the illustration at the left, 7 × 7 = 49, write 9 and carry 4; 6 + 4 = 10, 10 × 7 = 70, 70 + 4 = 74, write 4 and carry 7; 4 × 6 = 24, 24 + 7 = 31, write 31.

In the illustration at the right, 13 × 13 = 169, write 69 and carry 1; 6 + 9 = 15, 15 × 13 = 195, 195 + 1 = 196, write 96 and carry 1; 6 × 9 = 54, 54 + 1 = 55, write 55.

Perform the following multiplications by this method.

1. 136 × 56
2. 159 × 79
3. 172 × 92

4. 195 × 115
5. 234 × 174
6. 217 × 197

7. 516 × 816
8. 714 × 314
9. 217 × 917

Exercise No. 373
Multiplying Three Figures by Three

1. 152 × 909
2. 263 × 919
3. 374 × 929
4. 485 × 939

5. 596 × 949
6. 647 × 959
7. 758 × 969
8. 869 × 979

9. 973 × 989
10. 184 × 999

Exercise No. 374
Multiplying When Tens or Hundreds Are Alike

This is a variation of the method explained in Exercise No. 372 above.

```
    83              717
    89              714
  7387            511938
```

In the example on page 139, $3 \times 9 = 27$, write 7 and carry 2; $3 + 9 = 12$, $12 \times 8 = 96$, $96 + 2 = 98$, write 8 and carry 9; $8 \times 8 = 64$, $64 + 9 = 73$, write 73.

In the example on page 139, $17 \times 14 = 238$, write 38 and carry 2; $17 + 14 = 31$, $31 \times 7 = 217$, $217 + 2 = 219$, write 19 and carry 2; $7 \times 7 = 49$, $49 + 2 = 51$, write 51.

Multiply the following by this method.

1. 92×93	4. 92×97	7. 416×418
2. 62×65	5. 213×215	8. 509×519
3. 84×87	6. 321×312	9. 913×917

Exercise No. 375

Square of Numbers Ending in 5

If a number to be squared consists of tens and units, and if the units are 5, then twice the product of the first part by the second is equal to the given number of tens. Thus, in 25×25, $20 \times 5 \times 2$ is equal to 20×10; in 35×35, $30 \times 5 \times 2$ is equal to 30×10. Accordingly when dealing with numbers of this type we may at once annex 25 to the product of the given tens multiplied by one more than the given tens. That is to say, $25 \times 25 = 625$, in which the 6 represents 3×2; $35 \times 35 = 1225$ in which the 12 represents 4×3; $45 \times 45 = 2025$, in which the 20 represents 5×4, etc.

Find the squares of the following numbers by this method.

1. 45	4. 75	7. 115	10. 175	13. 335
2. 55	5. 85	8. 135	11. 195	14. 355
3. 65	6. 95	9. 155	12. 315	15. 375

Exercise No. 376

Multiplying Like Tens with Units Making 10

The principle explained above applies to any case in which the tens are alike and the sum of the units is 10.

Thus the product of 46 × 44 is 2024. We arrive at this by multiplying 4 × 5, making 20, and writing after this the product of 4 × 6 or 24.

Multiply in this manner the following.

1. 23 × 27	**4.** 103 × 107	**7.** 178 × 172
2. 41 × 49	**5.** 112 × 118	**8.** 169 × 161
3. 36 × 34	**6.** 154 × 156	**9.** 192 × 198

Exercise No. 377

Squaring Numbers Ending in 25

When a number ends in 25, like 725 for instance, we may take it as the sum of two numbers of which one represents hundreds and the other tens and units. In such cases twice the product of the first part by the second is equal to 50 times the first part. The result of this multiplication is a certain number of thousands.

To find the square of 725 we first write 0625 after the square of 7, making 490625. To this we add as many thousands as are represented by 7 × 5. 490625 + 35000 = 525625.

Another method of finding these squares is by setting the numbers down as in the following illustration.

$$
\begin{array}{r}
725 \\
725 \\
\hline
525625
\end{array}
$$

At once write 625 as the square of 25. Multiply 7 by 5, write 5 and carry 3; multiply 7 by 7, add 3, write 52.

Find the square of the following numbers by both of the foregoing methods.

1. 525	**3.** 825	**5.** 1225	**7.** 1625	**9.** 1825
2. 625	**4.** 1025	**6.** 1325	**8.** 1725	**10.** 1925

Exercise No. 378

Multiplying a Sum by a Difference

The algebraic product of $a + b$ and $a - b$ is $a^2 - b^2$. When numbers to be multiplied can be expressed as the sum of and the difference between two numbers, the product equals the square of the first minus the square of the second. Thus 63×57 may be expressed as $60 + 3$ multiplied by $60 - 3$. The product equals 60×60 minus 3×3. This comes to $3600 - 9$ or 3591.

There is no limit to the combinations of numbers for which this principle would hold true, but for practical purposes we may be satisfied to recognize those in which the units add to 10 and the tens have a difference of 1.

Multiply the following by this method.

1. 72×68	**4.** 101×119	**7.** 152×168
2. 83×77	**5.** 123×137	**8.** 173×187
3. 94×86	**6.** 146×154	**9.** 182×198

Exercise No. 379

Multiplying Mixed Numbers with Like Integers

When integers are alike in mixed numbers, as in $9\frac{1}{4} \times 9\frac{3}{4}$, their product is found by multiplying one integer by the other plus the sum of the two fractions; to this partial product add that obtained by multiplying together the two fractions.

$$
\begin{array}{cc}
9\frac{1}{4} & 8\frac{3}{4} \\
9\frac{3}{4} & 8\frac{5}{6} \\
\hline
90\frac{3}{16} & 76\frac{2}{3} \\
& \frac{5}{8} \\
\hline
& 77\frac{7}{24}
\end{array}
$$

In the illustrative example at the left, 9 is multiplied by $9 + \frac{1}{4} + \frac{3}{4}$, or 10. The product of this is 90, and to 90 is added the product of $\frac{1}{4}$ and $\frac{3}{4}$, or $\frac{3}{16}$.

In the second example 8 is multiplied by $8 + \frac{3}{4} + \frac{5}{6}$, or $9\frac{7}{12}$, producing $76\frac{2}{3}$. To this is added the product of $\frac{3}{4} \times \frac{5}{6}$, or $\frac{5}{8}$, making a total of $77\frac{7}{24}$.

Multiply the following.

1. $9\frac{1}{3} \times 9\frac{2}{3}$	**5.** $3\frac{1}{3} \times 3\frac{2}{3}$	**9.** $5\frac{1}{4} \times 5\frac{1}{2}$
2. $10\frac{3}{5} \times 10\frac{2}{5}$	**6.** $60\frac{3}{5} \times 60\frac{3}{4}$	**10.** $8\frac{3}{4} \times 8\frac{1}{3}$
3. $12\frac{5}{6} \times 12\frac{1}{2}$	**7.** $40\frac{3}{8} \times 40\frac{1}{4}$	**11.** $6\frac{5}{8} \times 6\frac{3}{8}$
4. $18\frac{1}{2} \times 18\frac{1}{3}$	**8.** $25\frac{3}{5} \times 25\frac{2}{5}$	**12.** $12\frac{1}{9} \times 12\frac{5}{9}$

Exercise No. 380

Multiplying by a Number Nearly Whole

Sometimes a multiplier lacks a single fractional unit of being a whole number. Examples would be $5\frac{2}{3}$, $6\frac{3}{4}$ and $7\frac{4}{5}$, which respectively lack $\frac{1}{3}$, $\frac{1}{4}$ and $\frac{1}{5}$ of being 6, 7 and 8. In cases of this kind raise the multiplier to the next larger whole number, and after multiplying the multiplicand by this number, subtract from the product the necessary fractional part of the multiplicand. Thus, to multiply 64 by $3\frac{7}{8}$, we multiply 64 by 4, obtaining 256, and from this we subtract $\frac{1}{8}$ of 64, or 8, arriving at a final result of 248.

Multiply by this method the following.

1. $48 \times 5\frac{3}{4}$	**4.** $250 \times 3\frac{4}{5}$	**7.** $180 \times 7\frac{9}{10}$
2. $75 \times 10\frac{2}{3}$	**5.** $522 \times 4\frac{8}{9}$	**8.** $720 \times 2\frac{11}{12}$
3. $136 \times 6\frac{5}{6}$	**6.** $672 \times 8\frac{6}{7}$	**9.** $342 \times 9\frac{5}{6}$

Exercise No. 381

Aliquot Parts in Division

The method of aliquot parts is as applicable to division as it is to multiplication. In ordinary cases we determine how many times the given divisor is contained exactly in some multiple of 10. We multiply the given dividend by the result of such division, and point off the product decimally in such a way as to express division by the proper multiple of 10. Thus, to divide 1840 by 25, we obtain a multiplier of 4 by dividing 25 into 100. Multiplying 1840 by 4 we get 7360, and dividing this decimally by 100 we obtain 73.60

$$6375 \div 7\tfrac{1}{2}$$

$$
\begin{array}{r}
6375 \\
2125 \\
\hline
850.0
\end{array}
$$

Another method of using aliquot parts is illustrated by the example shown above. The problem is to divide 6375 by $7\tfrac{1}{2}$. We note that $7\tfrac{1}{2}$ lacks one-third of itself of being 10. We therefore add one-third of itself to 6375 and divide the resulting sum decimally by 10.

Divide by the foregoing methods:

1. $580 \div 25$	**4.** $875 \div 250$	**7.** $1527 \div 150$
2. $750 \div 16\tfrac{2}{3}$	**5.** $640 \div 125$	**8.** $918 \div 15$
3. $450 \div 12\tfrac{1}{2}$	**6.** $435 \div 33\tfrac{1}{3}$	**9.** $582 \div 7\tfrac{1}{2}$

Exercise No. 382

Cubes of Numbers

The algebraic formula for the cube of the sum of two numbers, a and b, is $a^3 + 3a^2b + 3ab^2 + b^3$. This may be expressed as the cube of the first plus three times the square of the first multiplied by the second, plus three times the first multiplied by the square of the second plus the cube of the second.

By applying this formula it is not difficult to calculate mentally the cubes of numbers of two places. Suppose, for instance, that we want to find the cube of 26. We immediately annex the cube of 6 (216) to the cube of 2 (8), obtaining 8216. (Always allow three places for the cube of the second.) Multiplying 3×400 (square of 20) $\times 6$, we get 7200, which, added to 8216, makes 15416. Multiplying $3 \times 20 \times 36$ (square of 6) we obtain 2160, which, added to 15416 gives 17576 as the cube of 26.

Cubes may be readily written down from right to left by using a different method.

$$
\begin{array}{lll}
\dfrac{26^3}{17576} & 6 \times 6 \times 6 = 216 & 6 \\
 & (6 \times 6 \times 2 \times 3) + 21 = 237 & 7 \\
 & (6 \times 2 \times 2 \times 3) + 23 = 95 & 5 \\
 & (2 \times 2 \times 2) + 9 = 17 & 17
\end{array}
$$

All the necessary writing is shown on p.144 at the left. The method of making the calculation is analyzed at the right. The cube of 6 is 216, write 6 and carry 21. The square of 6 (36) multiplied by 2 (72) multiplied by 3 (216) plus 21 comes to 237, write 7 and carry 23. The product of 6 times the square of 2 (24) multiplied by 3 (72) plus 23 comes to 95, write 5 and carry 9. The cube of 2 is 8, which, added to 9, makes 17.

Before attempting the examples which follow the student ought to make himself thoroughly familiar with the cubes of the numbers from 1 to 9, so that he will not have to slow up to make such computations in the course of the example.

Find the cubes of the following numbers by both of the foregoing methods.

1. 14	**4.** 46	**7.** 65	**10.** 84	**13.** 95
2. 27	**5.** 59	**8.** 71	**11.** 86	**14.** 97
3. 33	**6.** 62	**9.** 73	**12.** 88	**15.** 99

Exercise No. 383
Algebraic Multiplication

Arithmetical products may be directly written down from right to left by using the method of cross-multiplication employed in algebra. A certain pattern is followed in multiplying each figure by every other figure. The operations are best explained by illustration.

47	345
26	678
1222	234910

In the example at the left, $7 \times 6 = 42$, write 2 and carry 4; 4 plus 4×6 (28) plus 2×7 comes to 42, write 2 and carry 4; 4 plus 4×2 is 12, write 12. (It is best to start each part of the calculation with the carried number, which otherwise might not be easy to remember.)

In the second example, multiply 5×8; then 4×8 and 7×5; then 3×8, 6×5 and 4×7; then 3×7 and 6×4; finally 3×6. Carry as may be necessary.

Table IV
Prime and Composite Numbers

1 Prime	41 Prime	71 Prime	98 = 2 × 49
2 Prime	42 = 2 × 21	72 = 2 × 36	7 × 14
3 Prime	3 × 14	3 × 24	99 = 3 × 33
4 = 2 × 2	6 × 7	4 × 18	9 × 11
5 Prime	43 Prime	6 × 12	100 = 2 × 50
6 = 2 × 3	44 = 2 × 22	8 × 9	4 × 25
7 Prime	4 × 11	73 Prime	5 × 20
8 = 2 × 4	45 = 3 × 15	74 = 2 × 37	10 × 10
9 = 3 × 3	5 × 9	75 = 3 × 25	101 Prime
10 = 2 × 5	46 = 2 × 23	5 × 15	102 = 2 × 51
11 Prime	47 Prime	76 = 2 × 38	3 × 34
12 = 2 × 6	48 = 2 × 24	4 × 19	6 × 17
3 × 4	3 × 16	77 = 7 × 11	103 Prime
13 Prime	4 × 12	78 = 2 × 39	104 = 2 × 52
14 = 2 × 7	6 × 8	3 × 26	4 × 26
15 = 3 × 5	49 = 7 × 7	6 × 13	8 × 13
16 = 2 × 8	50 = 2 × 25	79 Prime	105 = 3 × 35
4 = 4	5 × 10	80 = 2 × 40	5 × 21
17 Prime	51 = 3 × 17	4 × 20	7 × 15
18 = 2 × 9	52 = 2 × 26	5 × 16	106 = 2 × 53
3 × 6	4 × 13	8 × 10	107 Prime
19 Prime	53 Prime	81 = 3 × 27	108 = 2 × 54
20 = 2 × 10	54 = 2 × 27	9 × 9	3 × 36
4 × 5	3 × 18	82 = 2 × 41	4 × 27
21 = 3 × 7	6 × 9	83 Prime	6 × 18
22 = 2 × 11	55 = 5 × 11	84 = 2 × 42	9 × 12
23 Prime	56 = 2 × 28	3 × 28	109 Prime
24 = 2 × 12	4 × 14	4 × 21	110 = 2 × 55
3 × 8	7 × 8	6 × 14	5 × 22
4 × 6	57 = 3 × 19	7 × 12	10 × 11
25 = 5 × 5	58 = 2 × 29	85 = 5 × 17	111 = 3 × 37
26 = 2 × 13	59 Prime	86 = 2 × 43	112 = 2 × 56
27 = 3 × 9	60 = 2 × 30	87 = 3 × 29	4 × 28
28 = 2 × 14	3 × 20	88 = 2 × 44	7 × 16
4 × 7	4 × 15	4 × 22	8 × 14
29 Prime	5 × 12	8 × 11	113 Prime
30 = 2 × 15	6 × 10	89 Prime	114 = 2 × 57
3 × 10	61 Prime	90 = 2 × 45	3 × 38
5 × 6	62 = 2 × 31	3 × 30	6 × 19
31 Prime	63 = 3 × 21	5 × 18	115 = 5 × 23
32 = 2 × 16	7 × 9	6 × 15	116 = 2 × 58
4 × 8	64 = 2 × 32	9 × 10	4 × 29
33 = 3 × 11	4 × 16	91 = 7 × 13	117 = 3 × 39
34 = 2 × 17	8 × 8	92 = 2 × 46	9 × 13
35 = 5 × 7	65 = 5 × 13	4 × 23	118 = 2 × 59
36 = 2 × 18	66 = 2 × 33	93 = 3 × 31	119 = 7 × 17
3 × 12	3 × 22	94 = 2 × 47	120 = 2 × 60
4 × 9	6 × 11	95 = 5 × 19	3 × 40
6 × 6	67 Prime	96 = 2 × 48	4 × 30
37 Prime	68 = 2 × 34	3 × 32	5 × 24
38 = 2 × 19	4 × 17	4 × 24	6 × 20
39 = 3 × 13	69 = 3 × 23	6 × 16	8 × 15
40 = 2 × 20	70 = 2 × 35	8 × 12	10 × 12
4 × 10	5 × 14	97 Prime	121 = 11 × 11
5 × 8	7 × 10		122 = 2 × 61

Table IV (Continued)

123 = 3 × 41	149 Prime	173 Prime	196 = 2 × 98
124 = 2 × 62	150 = 2 × 75	174 = 2 × 87	4 × 49
4 × 31	3 × 50	3 × 58	7 × 28
125 = 5 × 25	5 × 30	6 × 29	14 × 14
126 = 2 × 63	6 × 25	175 = 5 × 35	197 Prime
3 × 42	10 × 15	7 × 25	198 = 2 × 99
6 × 21	151 Prime	176 = 2 × 88	3 × 66
7 × 18	152 = 2 × 76	4 × 44	6 × 33
9 × 14	4 × 38	8 × 22	9 × 22
127 Prime	8 × 19	11 × 16	11 × 18
128 = 2 × 64	153 = 3 × 51	177 = 3 × 59	199 Prime
4 × 32	9 × 17	178 = 2 × 89	200 = 2 × 100
8 × 16	154 = 2 × 77	179 Prime	4 × 50
129 = 3 × 43	7 × 22	180 = 2 × 90	5 × 40
130 = 2 × 65	11 × 14	3 × 60	8 × 25
5 × 26	155 = 5 × 31	4 × 45	10 × 20
10 × 13	156 = 2 × 78	5 × 36	201 = 3 × 67
131 Prime	3 × 52	6 × 30	202 = 2 × 101
132 = 2 × 66	4 × 39	9 × 20	203 = 7 × 29
3 × 44	6 × 26	10 × 18	204 = 2 × 102
4 × 33	12 × 13	12 × 15	3 × 68
6 × 22	157 Prime	181 Prime	4 × 51
11 × 12	158 = 2 × 79	182 = 2 × 91	6 × 34
133 = 7 × 19	159 = 3 × 53	7 × 26	12 × 17
134 = 2 × 67	160 = 2 × 80	13 × 14	205 = 5 × 41
135 = 3 × 45	4 × 40	183 = 3 × 61	206 = 2 × 103
5 × 27	5 × 32	184 = 2 × 92	207 = 3 × 69
9 × 15	8 × 20	4 × 46	9 × 23
136 = 2 × 68	10 × 16	8 × 23	208 = 2 × 104
4 × 34	161 = 7 × 23	185 = 5 × 37	4 × 52
8 × 17	162 = 2 × 81	186 = 2 × 93	8 × 26
137 Prime	3 × 54	3 × 62	13 × 16
138 = 2 × 69	6 × 27	6 × 31	209 = 11 × 19
3 × 46	9 × 18	187 = 11 × 17	210 = 2 × 105
6 × 23	163 Prime	188 = 2 × 94	3 × 70
139 Prime	164 = 2 × 82	4 × 47	5 × 42
140 = 2 × 70	4 × 41	189 = 3 × 63	6 × 35
4 × 35	165 = 3 × 55	7 × 27	7 × 30
5 × 28	5 × 33	9 × 21	10 × 21
7 × 20	11 × 15	190 = 2 × 95	14 × 15
10 × 14	166 = 2 × 83	5 × 38	211 Prime
141 = 3 × 47	167 Prime	10 × 19	212 = 2 × 106
142 = 2 × 71	168 = 2 × 84	191 Prime	4 × 53
143 = 11 × 13	3 × 56	192 = 2 × 96	213 = 3 × 71
144 = 2 × 72	4 × 42	3 × 64	214 = 2 × 107
3 × 48	6 × 28	4 × 48	215 = 5 × 43
4 × 36	7 × 24	6 × 32	216 = 2 × 108
6 × 24	8 × 21	8 × 24	3 × 72
8 × 18	12 × 14	12 × 16	4 × 54
9 × 16	169 = 13 × 13	193 Prime	6 × 36
12 × 12	170 = 2 × 85	194 = 2 × 97	8 × 27
145 = 5 × 29	5 × 34	195 = 3 × 65	9 × 24
146 = 2 × 73	10 × 17	5 × 39	12 × 18
147 = 3 × 49	171 = 3 × 57	13 × 15	217 = 7 × 31
7 × 21	9 × 19		218 = 2 × 109
148 = 2 × 74	172 = 2 × 86		219 = 3 × 73
4 × 37	4 × 43		

Table IV (Continued)

$220 = 2 \times 110$	$240 = 2 \times 120$	$261 = 3 \times 87$	283 Prime
4×55	3×80	9×29	$284 = 2 \times 142$
5×44	4×60	$262 = 2 \times 131$	4×71
10×22	5×48	263 Prime	$285 = 3 \times 95$
11×20	6×40	264 2×132	5×57
$221 = 13 \times 17$	8×30	3×88	15×19
$222 = 2 \times 111$	10×24	4×66	$286 = 2 \times 143$
3×74	12×20	6×44	11×26
6×37	15×16	8×33	13×22
223 Prime	241 Prime	11×24	$287 = 7 \times 41$
$224 = 2 \times 112$	$242 = 2 \times 121$	12×22	$288 = 2 \times 144$
4×56	11×22	$265 = 5 \times 53$	3×96
7×32	$243 = 3 \times 81$	$266 = 2 \times 133$	4×72
8×28	9×27	7×38	6×48
14×16	$244 = 2 \times 122$	14×19	8×36
$225 = 3 \times 75$	4×61	$267 = 3 \times 89$	9×32
5×45	$245 = 5 \times 49$	$268 = 2 \times 134$	12×24
9×25	7×35	4×67	16×18
15×15	$246 = 2 \times 123$	269 Prime	$289 = 17 \times 17$
$226 = 2 \times 113$	3×82	$270 = 2 \times 135$	$290 = 2 \times 145$
227 Prime	6×41	3×90	5×58
$228 = 2 \times 114$	$247 = 13 \times 19$	5×54	10×29
3×76	$248 = 2 \times 124$	6×45	$291 = 3 \times 97$
4×57	4×62	9×30	$292 = 2 \times 146$
6×38	8×31	10×27	4×73
12×19	$249 = 3 \times 83$	15×18	293 Prime
229 Prime	$250 = 2 \times 125$	271 Prime	$294 = 2 \times 147$
$230 = 2 \times 115$	5×50	$272 = 2 \times 136$	3×98
5×46	10×25	4×68	6×49
10×23	251 Prime	8×34	7×42
$231 = 3 \times 77$	$252 = 2 \times 126$	16×17	14×21
7×33	3×84	$273 = 3 \times 91$	$295 = 5 \times 59$
11×21	4×63	7×39	$296 = 2 \times 148$
$232 = 2 \times 116$	6×42	13×21	4×74
4×58	7×36	$274 = 2 \times 137$	8×37
8×29	9×28	$275 = 5 \times 55$	$297 = 3 \times 99$
233 Prime	12×21	11×25	9×33
$234 = 2 \times 117$	14×18	$276 = 2 \times 138$	11×27
3×78	$253 = 11 \times 23$	3×92	$298 = 2 \times 149$
6×39	$254 = 2 \times 127$	4×69	$299 = 13 \times 23$
9×26	$255 = 3 \times 85$	6×46	$300 = 2 \times 150$
13×18	5×51	12×23	3×100
$235 = 5 \times 47$	15×17	277 Prime	4×75
$236 = 2 \times 118$	$256 = 2 \times 128$	$278 = 2 \times 139$	5×60
4×59	4×64	$279 = 3 \times 93$	6×50
$237 = 3 \times 79$	8×32	9×31	10×30
$238 = 2 \times 119$	16×16	$280 = 2 \times 140$	12×25
7×34	257 Prime	4×70	15×20
14×17	$258 = 2 \times 129$	5×56	$301 = 7 \times 43$
239 Prime	3×86	7×40	$302 = 2 \times 151$
	6×43	8×35	$303 = 3 \times 101$
	$259 = 7 \times 37$	10×28	$304 = 2 \times 152$
	$260 = 2 \times 130$	14×20	4×76
	4×65	281 Prime	8×38
	5×52	$282 = 2 \times 141$	16×19
	10×26	3×94	$305 = 5 \times 61$
	13×20	6×47	

Table IV (Continued)

306 = 2 × 153	326 = 2 × 163	348 = 2 × 174	*368* = 2 × 184
3 × 102	327 = 3 × 109	3 × 116	4 × 92
6 × 51	328 = 2 × 164	4 × 87	8 × 46
9 × 34	4 × 82	6 × 58	16 × 23
17 × 18	8 × 41	12 × 29	369 = 3 × 123
307 Prime	329 = 7 × 47	349 Prime	9 × 41
308 = 2 × 154	*330* = 2 × 165	*350* = 2 × 175	370 = 2 × 185
4 × 77	3 × 110	5 × 70	5 × 74
7 × 44	5 × 66	7 × 50	10 × 37
11 × 28	6 × 55	10 × 35	371 = 5 × 53
14 × 22	10 × 33	14 × 25	372 = 2 × 186
309 = 3 × 103	11 × 30	351 = 3 × 117	3 × 124
310 = 2 × 155	15 × 22	9 × 39	4 × 93
5 × 62	331 Prime	13 × 27	6 × 62
10 × 31	332 = 2 × 166	*352* = 2 × 176	12 × 31
311 = Prime	4 × 83	4 × 88	373 Prime
312 = 2 × 156	333 = 3 × 111	8 × 44	*374* = 2 × 187
3 × 104	9 × 37	11 × 32	11 × 34
4 × 78	334 = 2 × 167	16 × 22	17 × 22
6 × 52	335 = 5 × 67	353 Prime	*375* = 3 × 125
8 × 39	*336* = 2 × 168	354 = 2 × 177	5 × 75
12 × 26	3 × 112	3 × 118	15 × 25
13 × 24	4 × 84	6 × 59	376 = 2 × 188
313 Prime	6 × 56	355 = 5 × 71	4 × 94
314 = 2 × 157	7 × 48	356 = 2 × 178	8 × 47
315 = 3 × 105	8 × 42	4 × 89	377 = 13 × 29
5 × 63	12 × 28	*357* = 3 × 119	*378* = 2 × 189
7 × 45	14 × 24	7 × 51	3 × 126
9 × 35	16 × 21	17 × 21	6 × 63
15 × 21	337 Prime	358 = 2 × 179	7 × 54
316 = 2 × 158	338 = 2 × 169	359 Prime	9 × 42
4 × 79	13 × 26	*360* = 2 × 180	14 × 27
317 Prime	339 = 3 × 113	3 × 120	18 × 21
318 = 2 × 159	*340* = 2 × 170	4 × 90	379 Prime
3 × 106	4 × 85	5 × 72	*380* = 2 × 190
6 × 53	5 × 68	6 × 60	4 × 95
319 = 11 × 29	10 × 34	8 × 45	5 × 76
320 = 2 × 160	17 × 20	9 × 40	10 × 38
4 × 80	341 = 11 × 31	10 × 36	19 × 20
5 × 64	*342* = 2 × 171	12 × 30	381 = 3 × 127
8 × 40	3 × 114	15 × 24	382 = 2 × 191
10 × 32	6 × 57	18 × 20	383 Prime
16 × 20	9 × 38	*361* = 19 × 19	*384* = 2 × 192
321 = 3 × 107	18 × 19	362 = 2 × 181	3 × 128
322 = 2 × 161	343 = 7 × 49	363 = 3 × 121	4 × 96
7 × 46	344 = 2 × 172	11 × 33	6 × 64
14 × 23	4 × 86	364 = 2 × 182	8 × 48
323 = 17 × 19	8 × 43	4 × 91	12 × 32
324 = 2 × 162	*345* = 3 × 115	7 × 52	16 × 24
3 × 108	5 × 69	13 × 28	385 = 5 × 77
4 × 81	15 × 23	14 × 26	7 × 55
6 × 54	346 = 2 × 173	365 = 5 × 73	11 × 35
9 × 36	347 Prime	366 = 2 × 183	386 = 2 × 193
12 × 27		3 × 122	387 = 3 × 129
18 × 18		6 × 61	9 × 43
325 = 5 × 65		367 Prime	388 = 2 × 194
13 × 25			4 × 97

Table IV (Continued)

389 Prime	408 = 2 × 204	429 = 3 × 143	448 = 2 × 224
390 = 2 × 195	3 × 136	11 × 39	4 × 112
3 × 130	4 × 102	13 × 33	7 × 64
5 × 78	6 × 68	430 = 2 × 215	8 × 56
6 × 65	8 × 51	5 × 86	14 × 32
10 × 39	12 × 34	10 × 43	16 × 28
13 × 30	17 × 24	431 Prime	449 Prime
15 × 26	409 Prime	432 = 2 × 216	450 = 2 × 225
391 = 17 × 23	410 = 2 × 205	3 × 144	3 × 150
392 = 2 × 196	5 × 82	4 × 108	5 × 90
4 × 98	10 × 41	6 × 72	6 × 75
7 × 56	411 = 3 × 137	8 × 54	9 × 50
8 × 49	412 = 2 × 206	9 × 48	10 × 45
14 × 28	4 × 103	12 × 36	15 × 30
393 = 3 × 131	413 = 7 × 59	16 × 27	18 × 25
394 = 2 × 197	414 = 2 × 207	18 × 24	451 = 11 × 41
395 = 5 × 79	3 × 138	433 Prime	452 = 2 × 226
396 = 2 × 198	6 × 69	434 = 2 × 217	4 × 113
3 × 132	9 × 46	7 × 62	453 = 3 × 151
4 × 99	18 × 23	14 × 31	454 = 2 × 227
6 × 66	415 = 5 × 83	435 = 3 × 145	455 = 5 × 91
9 × 44	416 = 2 × 208	5 × 87	7 × 65
11 × 36	4 × 104	15 × 29	13 × 35
12 × 33	8 × 52	436 = 2 × 218	456 = 2 × 228
18 × 22	13 × 32	4 × 109	3 × 152
397 Prime	16 × 26	437 = 19 × 23	4 × 114
398 = 2 × 199	417 = 3 × 139	438 = 2 × 219	6 × 76
399 = 3 × 133	418 = 2 × 109	3 × 146	8 × 57
7 × 57	11 × 38	6 × 73	12 × 38
19 × 21	19 × 22	439 Prime	19 × 24
400 = 2 × 200	419 Prime	440 = 2 × 220	457 Prime
4 × 100	420 = 2 × 210	4 × 110	458 = 2 × 229
5 × 80	3 × 140	5 × 88	459 = 3 × 153
8 × 50	4 × 105	8 × 55	9 × 51
10 × 40	5 × 84	10 × 44	17 × 27
16 × 25	6 × 70	11 × 40	460 = 2 × 230
20 × 20	7 × 60	20 × 22	4 × 115
401 Prime	10 × 42	441 = 3 × 147	5 × 92
402 = 2 × 201	12 × 35	7 × 63	10 × 46
3 × 134	14 × 30	9 × 49	20 × 23
6 × 67	15 × 28	21 × 21	461 Prime
403 = 13 × 31	20 × 21	442 = 2 × 221	462 = 2 × 231
404 = 2 × 202	421 Prime	13 × 34	3 × 154
4 × 101	422 = 2 × 211	17 × 26	6 × 77
405 = 3 × 135	423 = 3 × 141	443 Prime	7 × 66
5 × 81	9 × 47	444 = 2 × 222	11 × 42
9 × 45	424 = 2 × 212	3 × 148	14 × 33
15 × 27	4 × 106	4 × 111	21 × 22
406 = 2 × 203	8 × 53	6 × 74	463 Prime
7 × 58	425 = 5 × 85	12 × 37	464 = 2 × 232
14 × 29	17 × 25	445 = 5 × 89	4 × 116
407 = 11 × 37	426 = 2 × 213	446 = 2 × 223	8 × 58
	3 × 142	447 = 3 × 149	16 × 29
	6 × 71		465 = 3 × 155
	427 = 7 × 61		5 × 93
	428 = 2 × 214		15 × 31
	4 × 107		466 = 2 × 233

Table IV (Continued)

467 Prime	486 = 2 × 243	504 = 2 × 252	522 = 2 × 261
468 = 2 × 234	3 × 162	3 × 168	3 × 174
3 × 156	6 × 81	4 × 126	6 × 87
4 × 117	9 × 54	6 × 84	9 × 58
6 × 78	18 × 27	7 × 72	18 × 29
9 × 52	487 Prime	8 × 63	523 Prime
12 × 39	488 = 2 × 244	9 × 56	524 = 2 × 262
13 × 36	4 × 122	12 × 42	4 × 131
18 × 26	8 × 61	14 × 36	525 = 3 × 175
469 = 7 × 67	489 = 3 × 163	18 × 28	5 × 105
470 = 2 × 235	490 = 2 × 245	21 × 24	7 × 75
5 × 94	5 × 98	505 = 5 × 101	15 × 35
10 × 47	7 × 70	506 = 2 × 253	21 × 25
471 = 3 × 157	10 × 49	11 × 46	526 = 2 × 263
472 = 2 × 236	14 × 35	22 × 23	527 = 17 × 31
4 × 118	491 Prime	507 = 3 × 169	528 = 2 × 264
8 × 59	492 = 2 × 246	13 × 39	3 × 176
473 = 11 × 43	3 × 164	508 = 2 × 254	4 × 132
474 = 2 × 237	4 × 123	4 × 127	6 × 88
3 × 158	6 × 82	509 Prime	8 × 66
6 × 79	12 × 41	510 = 2 × 255	11 × 48
475 = 5 × 95	493 = 17 × 29	3 × 170	12 × 44
19 × 25	494 = 2 × 247	5 × 102	16 × 33
476 = 2 × 238	13 × 38	6 × 85	22 × 24
4 × 119	19 × 26	10 × 51	529 = 23 × 23
7 × 68	495 = 3 × 165	15 × 34	530 = 2 × 265
14 × 34	5 × 99	17 × 30	5 × 106
17 × 28	9 × 55	511 = 7 × 73	10 × 53
477 = 3 × 159	11 × 45	512 = 2 × 256	531 = 3 × 177
9 × 53	15 × 33	4 × 128	9 × 59
478 = 2 × 238	496 = 2 × 298	8 × 64	532 = 2 × 266
479 Prime	4 × 124	16 × 32	4 × 133
480 = 2 × 240	8 × 62	513 = 3 × 171	7 × 76
3 × 160	16 × 31	9 × 57	14 × 38
4 × 120	497 = 7 × 71	19 × 27	19 × 28
5 × 96	498 = 2 × 299	514 = 2 × 257	533 = 13 × 41
6 × 80	3 × 166	515 = 5 × 103	534 = 2 × 267
8 × 60	6 × 83	516 = 2 × 258	3 × 178
10 × 48	499 Prime	3 × 172	6 × 89
12 × 40	500 = 2 × 250	4 × 129	535 = 5 × 107
15 × 32	4 × 125	6 × 86	536 = 2 × 268
16 × 30	5 × 100	12 × 43	4 × 134
20 × 24	10 × 50	517 = 11 × 47	8 × 67
481 = 13 × 37	20 × 25	518 = 2 × 259	537 = 3 × 179
482 = 2 × 241	501 = 3 × 167	7 × 74	538 = 2 × 269
483 = 3 × 161	502 = 2 × 251	14 × 37	539 = 7 × 77
7 × 69	503 Prime	519 = 3 × 173	11 × 49
21 × 23		520 = 2 × 260	
484 = 2 × 242		4 × 130	
4 × 121		5 × 104	
11 × 44		8 × 65	
22 × 22		10 × 52	
485 = 5 × 97		13 × 40	
		20 × 26	
		521 Prime	

Table IV (Continued)

540 = 2 × 270	558 = 2 × 279	576 = 2 × 288	594 = 2 × 297
3 × 180	3 × 186	3 × 192	3 × 198
4 × 135	6 × 93	4 × 144	6 × 99
5 × 108	9 × 62	6 × 96	9 × 66
6 × 90	18 × 31	8 × 72	11 × 54
9 × 60	559 = 13 × 43	9 × 64	18 × 33
10 × 54	560 = 2 × 280	12 × 48	22 × 27
12 × 45	4 × 140	16 × 36	595 = 5 × 119
15 × 36	5 × 112	18 × 32	7 × 85
18 × 30	7 × 80	24 × 24	17 × 35
20 × 27	8 × 70	577 Prime	596 = 2 × 298
541 Prime	10 × 56	578 = 2 × 289	4 × 149
542 = 2 × 271	14 × 40	17 × 34	597 = 3 × 199
543 = 3 × 181	16 × 35	579 = 3 × 193	598 = 2 × 299
544 = 2 × 272	20 × 28	580 = 2 × 290	13 × 46
4 × 136	561 = 3 × 187	4 × 145	23 × 26
8 × 68	11 × 51	5 × 116	599 Prime
16 × 34	17 × 33	10 × 58	600 = 2 × 300
17 × 32	562 = 2 × 281	20 × 29	3 × 200
545 = 5 × 109	563 Prime	581 = 7 × 83	4 × 150
546 = 2 × 273	564 = 2 × 282	582 = 2 × 291	5 × 120
3 × 182	3 × 188	3 × 194	6 × 100
6 × 91	4 × 141	6 × 97	8 × 75
7 × 78	6 × 94	583 = 11 × 53	10 × 60
13 × 42	12 × 47	584 = 2 × 292	12 × 50
14 × 39	565 = 5 × 113	4 × 146	15 × 40
21 × 26	566 = 2 × 283	8 × 73	20 × 30
547 Prime	567 = 3 × 189	585 = 3 × 195	24 × 25
548 = 2 × 274	7 × 81	5 × 117	601 Prime
4 × 137	9 × 63	9 × 65	602 = 2 × 301
549 = 3 × 183	21 × 27	13 × 45	7 × 86
9 × 61	568 = 2 × 284	15 × 39	14 × 43
550 = 2 × 275	4 × 142	586 = 2 × 293	603 = 3 × 201
5 × 110	8 × 71	587 Prime	9 × 67
10 × 55	569 Prime	588 = 2 × 294	604 = 2 × 302
11 × 50	570 = 2 × 285	3 × 196	4 × 151
22 × 25	3 × 190	4 × 147	605 = 5 × 121
551 = 19 × 29	5 × 114	6 × 98	11 × 55
552 = 2 × 276	6 × 95	7 × 84	606 = 2 × 303
3 × 184	10 × 57	12 × 49	3 × 202
4 × 138	15 × 38	14 × 42	6 × 101
6 × 92	19 × 30	21 × 28	607 Prime
8 × 69	571 Prime	589 = 19 × 31	608 = 2 × 304
12 × 46	572 = 2 × 286	590 = 2 × 295	1 × 152
23 × 24	4 × 143	5 × 118	8 × 76
553 = 7 × 79	11 × 52	10 × 59	16 × 38
554 = 2 × 277	13 × 44	591 = 3 × 197	19 × 32
555 = 3 × 185	22 × 26	592 = 2 × 296	609 = 3 × 203
5 × 111	573 = 3 × 191	4 × 148	7 × 87
15 × 37	574 = 2 × 287	8 × 74	21 × 29
556 = 2 × 278	7 × 82	16 × 37	610 = 2 × 305
4 × 139	14 × 41	593 Prime	5 × 122
557 Prime	575 = 5 × 115		10 × 61
	23 × 25		611 = 13 × 47

Table IV (Concluded)

612 =	2 × 306	616 =	2 × 308	619	Prime	624 =	2 × 312
	3 × 204		4 × 154	620 =	2 × 310		3 × 208
	4 × 152		7 × 88		4 × 155		4× 156
	6 × 102		8 × 77		5 × 124		6 × 104
	9 × 68		11 × 56		10 × 62		8 × 78
	12 × 51		14 × 44		20 × 31		12 × 52
	17 × 36		22 × 28	621 =	3 × 207		13 × 48
	18 × 34	617	Prime		9 × 69		16 × 39
613	Prime	618 =	2 × 309		23 × 27		24 × 26
614 =	2 × 307		3 × 206	622 =	2 × 311	625 =	5 × 125
615 =	3 × 205		6 × 103	623 =	7 × 89		25 × 25
	5 × 123						
	15 × 41						

ANSWERS

The references at the head of each section are to the numbers of the exercises.

No. 1	30	70	69	53
	86	54	25	109
1. 32	42	110	81	65
2. 30	98	66	37	21
3. 29	26	22	93	77
4. 29	82	78	49	40
5. 29	38	34	105	96
6. 31	94	90	68	52
7. 31	50	53	24	108
8. 18	106	109	80	64
9. 37	62	65	36	48
10. 31	25	21	92	104
11. 25	81	77	20	60
12. 35	37	61	76	16
13. 34	93	17	32	72
14. 29	49	73	88	28
15. 26	105	29	44	84
16. 25	33	85	100	47
17. 30	89	41	56	103
18. 33	45	97	19	59
19. 27	101	60	75	15
20. 30	57	16	31	71
21. 33	13	72	87	55
22. 26	69	28	43	111
23. 28	32	84	99	67
24. 27	88		27	23
	44		83	79
	100	**No. 3**	39	35
No. 2	56		95	91
	40	1. 59	51	54
12	96	2. 51	107	110
68	52	3. 56	63	66
24	108	4. 70	26	22
80	64	5. 62	82	78
36	20	6. 55	38	62
92	76	7. 57	94	18
48	39	8. 59	50	74
104	95	9. 53	106	30
67	51	10. 51	34	86
23	107	11. 69	90	42
79	63	12. 58	46	98
35	47	13. 60	102	61
91	103	14. 65	58	17
19	59	15. 59	14	73
75	15	16. 61	70	29
31	71	17. 53	33	85
87	27	18. 53	89	
43	83		45	
99	46	**No. 4**	101	**No. 5**
55	102		57	
18	58	13	41	14
74	14		97	70

154

26	109	46	113	29
82	65	102	69	85
38	49	58	25	41
94	105	21	81	97
50	61	77	37	53
106	17	33	93	109
69	73	89	56	37
25	29	45	112	93
81	85	101	68	49
37	48	29	24	105
93	104	85	80	61
21	60	41	64	17
77	16	97	20	73
33	72	53	76	36
89	56	109	32	92
45	112	65	88	48
101	68	28	44	104
57	24	84	100	60
20	80	40	63	44
76	36	96	19	100
32	92	52	75	56
88	55	108	31	112
44	111	36	87	68
100	67	92	**No. 7**	24
28	23	48		80
84	79	104	16	43
40	63	60	72	99
96	19	16	28	55
52	75	72	84	111
108	31	35	40	67
64	87	91	96	51
27	43	47	52	107
83	99	103	108	63
39	62	59	71	19
95	18	43	27	75
51	74	99	83	31
107	30	55	39	87
35	86	111	95	50
91		67	23	106
47		23	79	62
103	**No. 6**	79	35	18
59		42	91	74
15	15	98	47	58
71	71	54	103	114
34	27	110	59	70
90	83	66	22	26
46	39	50	78	82
102	95	106	34	38
58	51	62	90	94
42	107	18	46	57
98	70	74	102	113
54	26	30	30	69
110	82	86	86	25
66	38	49	42	81
22	94	105	98	65
78	22	61	54	21
41	78	17	110	77
97	34	73	66	33
53	90	57		89

45	37	30	113	98
101	93	86	69	26
64	49	42	53	82
20	105	98	109	38
76	61	54	65	94
32	45	110	21	50
88	101	73	77	106
	57	29	33	62
No. 8	113	85	89	25
(Same as	69	41	52	81
No. 1)	25	97	108	37
	81	25	64	93
No. 9	44	81	20	49
17	100	37	76	105
73	56	93	60	33
29	112	49	116	89
85	68	105	72	45
41	52	61	28	101
97	108	24	84	57
53	64	80	40	113
109	20	36	96	69
72	76	92	59	32
28	32	48	115	88
84	88	104	71	44
40	51	32	27	100
96	107	88	83	56
24	63	44	67	112
80	19	100	23	40
36	75	56	79	96
92	59	112	35	52
48	115	68	91	108
104	71	31	47	64
60	27	87	103	20
23	83	43	66	76
79	39	99	22	39
35	95	55	78	95
91	58	111	34	51
47	114	39	90	107
103	70	95		63
31	26	51	**No. 11**	47
87	82	107		103
43	66	63	*(Same as*	59
99	22	19	*No. 3)*	115
55	78	75		71
111	34	38		27
67	90	94	**No. 12**	83
30	46	50		46
86	102	106	19	102
42	65	62	75	58
98	21	46	31	114
54	77	102	87	70
110	33	58	43	54
38	89	114	99	110
94		70	55	66
50		26	111	22
106	**No. 10**	82	74	78
62		45	30	34
18	18	101	86	90
74	74	57	42	53

Column 1

109
65
21
77
61
117
73
29
85
41
97
60
116
72
28
84
68
24
80
36
92
48
104
67
23
79
35
91

No. 13

1. 365
2. 268
3. 371
4. 433
5. 257
6. 327
7. 209
8. 270
9. 287
10. 410
11. 257
12. 404
13. 231
14. 217
15. 311
16. 303
17. 254
18. 237
19. 308
20. 343
21. 350
22. 360
23. 308
24. 271
25. 341

No. 14

20
76
32
88
44
100
56
112
75
31
87
43
99
27
83
39
95
51
107
63
26
82
38
94
50
106
34
90
46
102
58
114
70
33
89
45
101
57
113
41
97
53
109
65
21
77
40
96
52
108
64
48
104
60
116
72
28

Column 3

84
47
103
59
115
71
55
111
67
23
79
35
91
54
110
66
22
78
62
118
74
30
86
42
98
61
117
73
29
85
69
25
81
37
93
49
105
68
24
80
36
92

No. 15

1. 620
2. 777
3. 716
4. 562
5. 432
6. 590
7. 624
8. 716
9. 885
10. 828
11. 424
12. 592
13. 535

14. 656
15. 858

No. 16

21
77
33
89
45
101
57
113
76
32
88
44
100
28
84
40
96
52
108
64
27
83
39
95
51
107
35
91
47
103
59
115
71
34
90
46
102
58
114
42
98
54
110
66
22
78
41
97
53
109
65
49
105

Column 5

61
117
73
29
85
48
104
60
116
72
56
112
68
24
80
36
92
55
111
67
23
79
63
119
75
31
87
43
99
62
118
74
30
86
70
26
82
38
94
50
106
69
25
81
37
93

No. 17

1. 1059
2. 1055
3. 903
4. 963
5. 897
6. 1113
7. 1067
8. 759
9. 994

10. 932

No. 18

22
78
34
90
46
102
58
114
77
33
89
45
101
29
85
41
97
53
109
65
28
84
40
96
52
108
36
92
48
104
60
116
72
35
91
47
103
59
115
43
99
55
111
67
23
79
42
98
54
110
66
50
106
62

118
74
30
86
49
105
61
117
73
57
113
69
25
81
37
93
56
112
68
24
80
64
120
76
32
88
44
100
63
119
75
31
87
71
27
83
39
95
51
107
70
26
82
38
94

No. 19
1. 12
2. 34
3. 21
4. 56
5. 33
6. 78
7. 12
8. 13
9. 12
10. 21

11. 7
12. 34
13. 52
14. 11
15. 52

No. 20
1. 28
2. 28
3. 12
4. 19
5. 15
6. 26
7. 19
8. 18
9. 48
10. 21
11. 39
12. 17
13. 26
14. 58
15. 28
16. 18
17. 29
18. 19
19. 29

No. 21
23
79
35
91
47
103
59
115
78
34
90
46
102
30
86
42
98
54
110
66
29
85
41
97
53
109
37

93
49
105
61
117
73
36
92
48
104
60
116
44
100
56
112
68
24
80
43
99
55
111
67
51
107
63
119
75
31
87
50
106
62
118
74
58
114
70
26
82
38
94
57
113
69
25
81
65
121
77
33
89
45
101
64
120
76
32

88
72
28
84
40
96
52
108
71
27
83
39
95

No. 22

1. 294
2. 234
3. 414
4. 358
5. 379
6. 381
7. 370
8. 347
9. 221
10. 374

No. 23

1. 521
2. 213
3. 233
4. 321
5. 331
6. 313
7. 252
8. 412
9. 212
10. 130
11. 122
12. 441
13. 432
14. 351
15. 221

No. 24

24
80
36
92
48

104	115	31	91	**22.** 437
60	71	87	47	**23.** 722
116	27	43	103	**24.** 109
79	83	99	66	**25.** 515
35	39	55	122	**26.** 209
91	95	111	78	**27.** 336
47	58	39	34	**28.** 107
103	114	95	90	**29.** 868
31	70	51	74	**30.** 419
87	26	107	30	
43	82	63	86	
99	66	119	42	**No. 28**
55	122	75	98	
111	78	38	54	26
67	34	94	110	82
30	90	50	73	38
86	46	106	29	94
42	102	62	85	50
98	65	118	41	106
54	121	46	97	62
110	77	102		118
38	33	58		81
94	89	114		37
50	73	70	**No. 26**	93
106	29	26		49
62	85	82	**1.** $655.71	105
118	41	45	**2.** $751.32	33
74	97	101	**3.** $604.24	89
37	53	57	**4.** $577.21	45
93	109	113	**5.** $718.69	101
49	72	69	**6.** $769.64	57
105	28	53	**7.** $488.04	113
61	84	109	**8.** $691.93	69
117	40	65		32
45	96	121		88
101		77	**No. 27**	44
57	**No. 25**	33		100
113		89	**1.** 215	56
69	25	52	**2.** 415	112
25	81	108	**3.** 209	40
81	37	64	**4.** 329	96
44	93	120	**5.** 778	52
100	49	76	**6.** 109	108
56	105	60	**7.** 214	64
112	61	116	**8.** 248	120
68	117	72	**9.** 128	76
52	80	28	**10.** 237	39
108	36	84	**11.** 403	95
64	92	40	**12.** 106	51
120	48	96	**13.** 125	107
76	104	59	**14.** 125	63
32	32	115	**15.** 136	119
88	88	71	**16.** 204	47
51	44	27	**17.** 109	103
107	100	83	**18.** 143	59
63	56	67	**19.** 107	115
119	112	123	**20.** 308	71
75	68	79	**21.** 309	27
59		35		

83	83	110	35	118
46	39	66	91	74
102	95	122	47	30
58	51	78	103	86
114	107	62	59	70
70	63	118	115	126
54	119	74	71	82
110	82	30	34	38
66	38	86	90	94
122	94	42	46	50
78	50	98	102	106
34	106	61	58	69
90	34	117	114	125
53	90	73	42	81
109	46	29	98	37
65	102	85	54	93
121	58	69	110	79
77	114	125	66	33
61	70	81	112	89
117	33	37	78	45
73	89	93	41	101
29	45	49	97	57
85	101	105	53	113
41	57	68	109	76
97	113	124	65	32
60	41	80	121	88
116	97	36	49	44
72	53	92	105	100
28	109	76	61	
84	65	32	117	
68	121	88	73	
124	77	44	29	No. 31
80	40	100	85	
36	96	56	48	1. 621
92	52	112	104	2. 585
48	108	75	60	3. 687
104	54	31	116	4. 647
67	120	86	72	5. 630
123	48	43	56	6. 605
79	104	99	112	7. 570
35	60		68	8. 671
91	116		124	9. 625
75	72		80	10. 624
31	28	No. 30	36	
87	84		92	
43	47	28	55	
99	103	84	111	No. 32
55	59	40	67	
111	115	96	123	1. 161
74	71	52	79	2. 292
30	55	108	63	3. 71
86	111	64	119	4. 191
42	67	120	75	5. 171
98	123	83	31	6. 64
	79	39	87	7. 252
No. 29	35	95	43	8. 197
	91	51	99	9. 623
27	64	107	62	10. 284

ANSWERS 161

11. 94
12. 387
13. 170
14. 61
15. 593
16. 195
17. 394
18. 295
19. 492
20. 681

No. 33

1. 465
2. 579
3. 164
4. 186
5. 153
6. 48
7. 489
8. 186
9. 488
10. 377
11. 329
12. 469
13. 288
14. 56
15. 216
16. 184
17. 249
18. 77
19. 289
20. 169

No. 34

1. $995.69
2. $1044.85
3. $954.07
4. $1002.63
5. $994.32
6. $897.80
7. $1122.66
8. $1051.42

No. 35

1. 395
2. 297
3. 92
4. 299
5. 298
6. 195
7. 298
8. 399
9. 494

10. 497
11. 296
12. 94
13. 495
14. 294
15. 299
16. 198
17. 197
18. 397
19. 293
20. 692
21. 198
22. 294
23. 596
24. 99
25. 395

No. 36

1. 985
2. 987
3. 975
4. 1008
5. 953
6. 1011
7. 1042
8. 1032
9. 1095
10. 1012

No. 37

1. 347
2. 189
3. 349
4. 78
5. 107
6. 259
7. 189
8. 119
9. 66
10. 88
11. 215
12. 178
13. 178
14. 9
15. 227
16. 109
17. 114
18. 249
19. 234
20. 29
21. 298
22. 284
23. 38
24. 376
25. 129

No. 38

1. $42357.49
2. $57112.34
3. $54738.19
4. $62369.15
5. $70468.35
6. $63801.69

No. 39

1. $4.35
2. $5.59
3. $ .94
4. $1.48
5. $6.92
6. $7.63
7. $2.31
8. $6.84
9. $3.70
10. $2.76
11. $2.29
12. $6.76
13. $3.59
14. $5.96
15. $1.56
16. $3.89
17. $2.68
18. $6.92
19. $3.49
20. $5.97

No. 40

(Same as No. 13)

No. 41

1. $95513.02
2. $102635.78
3. $98506.46
4. $117398.69
5. $95153.78
6. $99073.91

No. 42

(Same as No. 39)

No. 43

1. $ .93
2. $1.20

3. $2.81
4. $ .65
5. $1.96
6. $5.84
7. $2.95
8. $1.65
9. $2.24
10. $ .71
11. $1.89
12. $ .73
13. $1.23
14. $1.63
15. $1.71
16. $2.48
17. $1.86
18. $1.94
19. $2.45
20. $1.63

No. 44

(Same as No. 43)

No. 45

2
114
26
138
50
162
74
186
112
24
136
48
160
16
128
40
152
64
176
88
14
126
38
150
62
174
30
142
54
166
78

190	124	174	228	336
102	36	63	52	160
28	148	231	276	384
140	60	99	100	208
52	172	267	324	144
164	98	135	148	368
76	10	87	372	192
188	122	255	224	16
44	34	123	48	240
156	146	291	272	64
68		159	96	288
180		27	320	140
92	**No. 46**	195	32	364
4		84	256	188
116	3	252	80	12
42	171	120	304	236
154	39	288	128	172
66	207	156	352	396
178	75	108	176	220
90	243	276	28	44
58	111	144	252	268
170	279	12	76	92
82	168	180	300	316
194	36	48	124	168
106	204	216	348	392
18	72	105	60	216
130	240	273	284	40
56	24	141	108	264
168	192	9	332	200
80	60	177	156	24
192	228	129	380	248
104	96	297	204	72
72	264	165	56	292
184	132	33	280	120
96	21	201	104	344
8	189	69	328	196
120	57	237	152	20
32	225	126	376	244
144	93	294	88	68
70	261	162	312	296
182	45	30	136	
94	213	198	360	
6	81	150	184	**No. 48**
118	249	18	8	
86	117	186	232	1. $3433540.07
198	285	54	84	2. $2509179.07
110	153	222	308	3. $3688667.60
22	42	90	132	4. $3251326.81
134	210	258	356	5. $3449296.55
46	78	147	180	6. $3353169.99
158	246	15	116	
84	114	183	340	
196	282	51	164	**No. 49**
108	66	219	388	
20	234		212	1. $18.53
132	102	**No. 47**	36	2. $25.66
100	270		260	3. $23.95
12	138	4	112	4. $14.78
	6			

5. $41.76	170	**No. 51**	174	259
6. $38.38	450		510	651
7. $15.74	230	*(Same as*	246	392
8. $42.95	10	*No. 49)*	582	84
9. $60.76	290		318	476
10. $71.19	105	**No. 52**	54	168
11. $66.57	385		390	560
12. $59.85	165	6	168	56
13. $93.72	445	342	504	448
14. $80.90	225	78	240	140
15. $75.68	145	414	576	532
16. $61.52	425	150	312	224
	205	486	216	616
	485	222	552	308
	265	558	288	49
No. 50	45	336	24	441
	325	72	360	133
5	140	408	96	525
285	420	144	432	217
65	200	480	210	609
345	480	48	546	105
125	260	384	282	497
405	180	120	18	189
185	460	456	354	581
465	240	192	258	273
280	20	528	594	665
60	300	264	330	357
340	80	42	66	98
120	360	378	402	490
400	175	114	138	182
40	455	450	474	574
320	235	186	252	266
100	15	522	588	658
380	295	90	324	154
160	215	426	60	546
440	495	162	396	238
220	275	498	300	630
35	55	234	36	322
315	335	570	372	14
95	115	306	108	406
375	395	84	444	147
155	210	420	180	539
435	490	156	516	231
75	270	492	294	623
355	50	228	30	315
135	330	564	366	203
415	250	132	102	595
195	30	468	438	287
475	310	204		679
255	90	540		371
70	370	276	**No. 53**	63
350	150	12		455
130	430	348	7	196
410	245	126	399	588
190	25	462	91	280
470	305	198	483	672
110	85	534	175	364
390	365	270	567	252

644	**12.** $55.60	712	**No. 59**	639
336	**13.** $97.15	360		243
28	**14.** $73.69	232	**1.** 795	747
420	**15.** $61.63	680	**2.** 682	351
112	**16.** $68.20	328	**3.** 564	855
504		776	**4.** 814	459
245		424	**5.** 598	126
637	**No. 56**	72	**6.** 924	630
329		520	**7.** 810	234
21	8	224	**8.** 946	738
413	456	672	**9.** 1032	342
301	104	320	**10.** 912	846
693	552	768	**11.** 901	198
385	200	416	**12.** 621	702
77	648	288	**13.** 665	306
469	296	736	**14.** 308	810
161	744	384	**15.** 962	414
553	448	32	**16.** 714	18
294	96	480	**17.** 1008	522
686	544	128	**18.** 364	189
378	192	576	**19.** 736	693
70	640	280	**20.** 782	297
462	64	728	**21.** 855	801
350	512	376	**22.** 864	405
42	160	24	**23.** 865	261
434	608	472	**24.** 988	765
126	256	344	**25.** 667	369
518	704	792		873
210	352	440		477
602	56	88	**No. 60**	81
343	504	536		585
35	152	184	9	252
427	600	632	513	756
119	248	336	117	360
511	696	784	621	864
	120	432	225	468
	568	80	729	324
No. 54	216	528	333	828
	664	400	837	432
1. $6537136.94	312	48	504	36
2. $6295852.28	760	496	108	540
3. $6328194.91	408	144	612	144
4. $5945296.77	112	592	216	648
	560	240	720	315
	208	688	72	819
No. 55	656	392	57 2	423
	304	40	180	27
1. $19.76	752	488	**684**	531
2. $18.86	176	136	288	387
3. $44.51	624	584	792	891
4. $26.39	272		396	495
5. $41.42	720	**No. 57**	63	99
6. $6.20	368	(*Same as*	567	603
7. $12.22	16	*No. 15*)	171	207
8. $19.63	464		675	711
9. $87.27	168	**No. 58**	279	378
10. $84.51	616	(*Same as*	783	882
11. $71.61	264	*No. 55*)	135	486

90	374	**No. 62**	**2.** $836.87
594	990		**3.** $666.99
450	506	**1.** $11230083.55	**4.** $829.97
54	22	**2.** $10797546.08	**5.** $634.22
558	608	**3.** $8876665.99	**6.** $827.43
162	231	**4.** $8230948.08	**7.** $857.76
666	847		**8.** $527.72
270	363		**9.** $418.44
774	979	**No. 63**	**10.** $906.92
441	495		**11.** $447.71
45	319	**1.** $47.65	**12.** $586.87
549	935	**2.** $6.21	**13.** $407.46
153	451	**3.** $79.61	**14.** $510.63
657	1067	**4.** $34.74	**15.** $533.62
	583	**5.** $14.68	**16.** $663.85
	99	**6.** $27.74	
No. 61	715	**7.** $27.93	
11	308	**8.** $21.85	**No. 68**
627	924	**9.** $54.46	
143	440	**10.** $13.83	(*Same as No. 17*)
759	1056	**11.** $36.49	
275	572	**12.** $4.46	
891	396	**13.** $50.47	**No. 69**
407	1012	**14.** $8.53	
1023	528	**15.** $27.16	(*Same as No. 67*)
616	44	**16.** $39.87	
132	660		
748	176		**No. 71**
264	792	**No. 65**	
880	385		**1.** $276.69
88	1001	(*Same as No. 63*)	**2.** $855.51
704	517		**3.** $682.90
220	33		**4.** $520.36
836	649	**No. 66**	**5.** $773.79
352	473		**6.** $891.54
968	1089	**1.** 1827	**7.** $326.93
484	605	**2.** 1705	**8.** $245.59
77	121	**3.** 1170	**9.** $371.93
693	737	**4.** 1376	**10.** $471.54
209	253	**5.** 2511	**11.** $386.88
825	869	**6.** 2624	**12.** $330.44
341	462	**7.** 3772	**13.** $878.62
957	1078	**8.** 1200	**14.** $696.89
165	594	**9.** 1537	**15.** $770.20
781	110	**10.** 1235	**16.** $674.87
297	726	**11.** 1408	
913	550	**12.** 1428	
429	66	**13.** 1407	**No. 72**
1045	682	**14.** 1408	
561	198	**15.** 2016	(*Same as No. 22*)
154	814	**16.** 2418	
770	330	**17.** 3772	
286	946	**18.** 1164	**No. 73**
902	539	**19.** 2015	
418	55	**20.** 2592	**1.** 755717535
1034	671		**2.** 756410013
242	187	**No. 67**	**3.** 824293224
858	803		**4.** 824985702
		1. $846.98	

5. 3674994324
6. 1167178458
7. 1236433047
8. 6091457406
9. 1690209807
10. 1752668607
11. 1511041308
12. 3675686802
13. 1306128921
14. 1031412036
15. 1442533509

No. 74

1. 1536
2. 4606
3. 2646
4. 1495
5. 5313
6. 3230
7. 7347
8. 4814
9. 4284
10. 1295
11. 6624
12. 1624
13. 1886
14. 3618
15. 5494
16. 3861
17. 3344
18. 8608
19. 1612
20. 2655

No. 75

(*Same as No. 71*)

No. 76

(*Same as No. 26*)

No. 77

12
684
156
828
300
972
444
1116
672

144
816
288
960
96
768
240
912
384
1056
528
84
756
228
900
372
1044
180
852
324
996
468
1140
612
168
840
312
984
456
1128
264
936
408
1080
552
24
696
252
924
396
1068
540
348
1020
492
1164
636
108
780
336
1008
480
1152
624
432
1104
576
48

720
192
864
420
1092
564
36
708
516
1188
660
132
804
276
948
504
1176
648
120
792
600
72
744
216
888
360
1032
588
60
732
204
876

No. 78

(*Same as No. 34*)

No. 79

1. $451.84
2. $189.86
3. $343.97
4. $352.59
5. $188.21
6. $145.71
7. $291.97
8. $664.63
9. $136.68
10. $86.14
11. $440.45
12. $221.48
13. $196.63
14. $146.23
15. $586.21
16. $568.49

No. 80

1. 17081

2. 13361
3. 25543
4. 22632
5. 37893
6. 34323
7. 52643
8. 45201
9. 68302
10. 62693
11. 19602
12. 12312
13. 77922
14. 33033
15. 25662
16. 12831
17. 16086
18. 20274
19. 22263
20. 47583
21. 44896

No. 81

1. 123782280
2. 123895704
3. 135014592
4. 135128016
5. 601943392
6. 191177264
7. 202520776
8. 997746448
9. 276846856
10. 287077256
11. 247500064
12. 602056816
13. 213936568
14. 168939488
15. 236278872

No. 82

(*Same as No. 38*)

No. 83

1. $451.84
2. $189.86
3. $343.97
4. $352.59
5. $188.21
6. $145.71
7. $291.97
8. $664.63
9. $136.68
10. $86.14
11. $440.45

12. $221.48	**15.** 256620	1001	**No. 93**
13. $196.63	**16.** 128310	429	
14. $146.23	**17.** 160860	1157	**1.** 195840
15. $586.21	**18.** 202740	585	**2.** 237930
16. $568.49	**19.** 222630	377	**3.** 282880
	20. 465830	1105	**4.** 244660
	21. 448960	533	**5.** 173440
No. 84		1261	**6.** 214830
		689	**7.** 242080
1. 19584	**No. 90**	117	**8.** 213460
2. 23793		845	**9.** 251640
3. 28288	13	364	**10.** 126910
4. 24466	741	1092	**11.** 171380
5. 17344	169	520	**12.** 219180
6. 21483	897	1248	**13.** 307020
7. 24208	325	676	**14.** 362060
8. 21346	1053	468	**15.** 333550
9. 25164	481	1196	**16.** 171990
10. 12691	1209	624	**17.** 278460
11. 17138	728	52	**18.** 310030
12. 21918	156	780	**19.** 291200
13. 30702	884	208	**20.** 339480
14. 36206	312	936	**21.** 162380
15. 33355	1040	455	
16. 17199	104	1183	
17. 27846	832	611	**No. 94**
18. 31003	260	39	
19. 29120	988	767	**1.** 135025095
20. 33948	416	559	**2.** 135148821
21. 16238	1144	1287	**3.** 147277608
	572	715	**4.** 147401334
	91	143	**5.** 656616308
	819	871	**6.** 208541386
No. 86	247	299	**7.** 220915199
	975	1027	**8.** 1088369102
1. $95513.02	403	546	**9.** 301992119
2. $102635.78	1131	1274	**10.** 303151719
3. $98506.46	195	702	**11.** 269979836
4. $117398.69	923	130	**12.** 656740034
5. $95153.78	351	858	**13.** 233367857
6. $99073.91	1079	650	**14.** 184383812
	507	78	**15.** 257739453
	1235	806	
	663	234	
No. 89	182	962	**No. 95**
	910	390	
1. 170810	338	1118	*(Same as No. 54)*
2. 133610	1066	637	
3. 255430	494	65	**No. 97**
4. 226320	1222	793	
5. 378930	286	221	**1.** 11211
6. 343230	1014	949	**2.** 24642
7. 526430	442		**3.** 40051
8. 452010	1170		**4.** 57902
9. 683020	598		**5.** 77691
10. 626930	26	**No. 91**	**6.** 92412
11. 196020	754		**7.** 29432
12. 123120	273	*(Same as No. 48)*	
13. 779220			
14. 330330			

8. 21311	**9.** 287	952	224
9. 35742	**10.** 410	336	1008
10. 52151	**11.** 257	1120	490
11. 71002	**12.** 404	112	1274
12. 91791	**13.** 231	896	658
13. 25521	**14.** 217	280	42
14. 48155	**15.** 311	1064	826
15. 24442	**16.** 303	448	602
16. 49184	**17.** 254	1232	1386
17. 76146	**18.** 237	616	770
18. 44844	**19.** 308	98	154
19. 37296	**20.** 343	882	938
20. 97902	**21.** 350	266	322
21. 39693	**22.** 360	1050	1106
	23. 308	434	588

No. 99

1. $11230083.55
2. $10797546.08
3. $8876665.99
4. $8230948.08

24. 271	1218
25. 341	210

No. 101

1. 36156
2. 59290
3. 80618
4. 22869
5. 36696
6. 52624
7. 71918
8. 93555
9. 97856
10. 103972
11. 108988
12. 84058
13. 103474
14. 108580
15. 79165
16. 57318
17. 65778
18. 77744
19. 91086
20. 35547
21. 80690

No. 103

1. 365
2. 268
3. 371
4. 433
5. 257
6. 327
7. 209
8. 270

No. 105

1. 116081
2. 142272
3. 165481
4. 107512
5. 132181
6. 159372
7. 156996
8. 191522
9. 181692
10. 217894
11. 110564
12. 110940
13. 121598
14. 120273
15. 134316
16. 120990
17. 113970
18. 145262
19. 122811
20. 139635
21. 144284

No. 106

14
798
182
966
350
1134
518
1302
784
168

1372
756
140
924
700
84
868
252
1036
420
1204
686
70
854
238
1022

994
378
1162
546
1330
714
196
980
364
1148
532
1316
308
1092
476
1260
644
28
812
294
1078
462
1246
630
406
1190
574
1358
742
126
910
392
1176
560
1344
728
504
1288
672
56
840

No. 107

(*Same as No. 17*)

No. 109

1. 136004
2. 229024
3. 268746
4. 128064
5. 160446
6. 236496
7. 195853
8. 223096
9. 368063
10. 145673
11. 187146
12. 305283
13. 355096
14. 291014
15. 348928
16. 145728
17. 336414
18. 395324

19. 430265
20. 247275
21. 575276

No. 110

1. 146267910
2. 146401938
3. 159540624
4. 159674652
5. 711289224
6. 225905508
7. 239309622
8. 1178991756
9. 327137382
10. 339226182
11. 292459608
12. 711423252
13. 252799146
14. 199628136
15. 279200034

No. 111

(*Same as No. 26*)

No. 113

1. 164232
2. 227238
3. 301464
4. 377910
5. 456576
6. 497502
7. 658752
8. 172104
9. 243320
10. 279396
11. 354252
12. 427652
13. 484432
14. 588078
15. 671944
16. 175392
17. 173514
18. 257237
19. 341968
20. 429525
21. 519302

No. 115

(*Same as No. 34*)

No. 118

(*Same as No. 38*)

No. 119

15
855
195
1035
375
1215
555
1395
840
180
1020
360
1200
120
960
300
1140
480
1320
660
105
945
285
1125
465
1305
225
1065
405
1245
585
1425
765
210
1050
390
1230
570
1410
330
1170
510
1350
690
30
870
315
1155
495
1335
675

435
1275
615
1455
795
135
975
420
1260
600
1440
780
540
1380
720
60
900
240
1080
525
1365
705
45
885
645
1485
825
165
1005
345
1185
630
1470
810
150
990
750
90
930
270
1110
450
1290
735
75
915
255
1095

No. 120

(*Same as No. 41*)

No. 122

(*Same as No. 48*)

No. 123

1. 157510725
2. 157655055
3. 171803640
4. 171947970
5. 765962140
6. 243269630
7. 257704045
8. 1269714410
9. 352282645
10. 365300645
11. 314939380
12. 766106470
13. 272230435
14. 214972460
15. 300660615

No. 124

(*Same as No. 54*)

No. 126

(*Same as No. 62*)

No. 128

(*Same as No. 38*)

No. 131

16
912
208
1104
400
1296
92
1488
896
192
1088
384
1280
128
1024
320
1216
512
1408
704
112
1008

304	368	340	51
1200	1264	1292	1003
496	672	544	731
1392	1568	1496	1683
240	864	748	935
1136	160	119	187
432	1056	1071	1139
1328	800	323	391
624	96	1275	1343
1520	992	527	714
816	288	1479	1666
224	1184	255	918
1120	480	1207	170
416	1376	459	1122
1312	784	1411	850
608	80	663	102
1504	976	1615	1054
352	272	867	306
1248	1168	238	1258
544		1190	510
1440		442	1462
736	**No. 132**	1394	833
32		646	85
928		1598	1037
336	**1.** 168753540	374	289
1232	**2.** 168908172	1326	1241
528	**3.** 184066656	578	
1424	**4.** 184221288	1530	
720	**5.** 820635056	782	**No. 141**
464	**6.** 260633752	34	
1360	**7.** 276098468	996	**1.** 179996355
656	**8.** 1360237064	357	**2.** 180161289
1552	**9.** 377427908	1309	**3.** 196329672
848	**10.** 391375108	561	**4.** 196494606
144	**11.** 337419152	1513	**5.** 875307972
1040	**12.** 820789688	765	**6.** 277997874
448	**13.** 291661724	493	**7.** 294492891
1344	**14.** 230316784	1445	**8.** 1450859718
640	**15.** 322121196	697	**9.** 402573171
1536		1649	**10.** 417449571
832		901	**11.** 359898924
576	**No. 140**	153	**12.** 875472906
1472		1105	**13.** 311093013
768	17	476	**14.** 245661108
64	969	1428	**15.** 343581777
960	221	680	
256	1173	1632	
1152	425	884	**No. 148**
560	1377	912	
1456	629	1564	18
752	1581	816	1026
48	952	68	234
944	204	1020	1242
688	1156	272	450
1584	408	1224	1458
880	1360	595	666
176	136	1547	1674
1072	1088	799	1008

216	1080	247	760
1224	288	1311	1824
432	1296	475	988
1440	630	1539	684
144	1638	703	1748
1152	846	1767	912
360	54	1064	76
1368	1062	228	1140
576	774	1292	304
1584	1782	456	1368
792	990	1520	665
126	198	152	1729
1134	1206	1216	893
342	414	380	57
1350	1422	1444	1121
558	756	608	817
1566	1764	1672	1881
270	972	836	1045
1278	180	133	209
486	1188	1197	1273
1494	900	361	437
702	108	1425	1501
1710	1116	589	798
918	324	1653	1862
252	1332	285	1026
1260	540	1349	190
468	1548	513	1254
1476	882	1577	950
684	90	741	114
1692	1098	1805	1178
396	306	969	342
1404	1314	266	1406
612		1330	570
1620		494	1634
828		1558	931
36		722	95
1044		1786	1159
378		418	323
1386		1482	1387
594		646	
1602		1710	
810		874	
522		38	
1530		1102	
738		399	
1746		1463	
954		627	
162		1691	
1170		855	
504		551	
1512		1615	
720		779	
1728		1843	
936		1007	
648		171	
1656		1235	
864		532	
72		1596	

No. 149

1. 191239170
2. 191414406
3. 208592688
4. 208767924
5. 929980808
6. 295361996
7. 312887314
8. 1541482372
9. 427718434
10. 443524034
11. 382378696
12. 930156124
13. 330524302
14. 261005432
15. 365042358

No. 156

19
1083

No. 159

1. 202481985
2. 202667523
3. 220855704
4. 221041242
5. 984653804
6. 312726118
7. 331281737
8. 1632105026
9. 452863697
10. 469598497
11. 404858468
12. 984839342
13. 349955591
14. 276349756
15. 386502939

No. 165	180	**13.** 369386880	1785
	1300	**14.** 291694080	861
20	560	**15.** 407963520	2037
1140	1680		1113
260	800	No. 172	189
1380	1920		1365
500	1040	21	588
1620	720	1197	1744
740	1840	273	840
1860	960	1449	2016
1120	80	525	1092
240	1200	1701	756
1360	320	777	1932
480	1440	1953	1008
1600	700	1176	84
160	1820	252	1260
1280	940	1428	336
400	60	504	1512
1520	1180	1680	735
640	860	168	1911
1760	1980	1344	987
880	1100	420	63
140	220	1596	1239
1260	1340	672	903
380	460	1848	2079
1500	1580	924	1155
620	840	147	231
1740	1960	1323	1407
300	1080	399	483
1420	200	1575	1659
540	1320	651	882
1660	1000	1827	2058
780	120	315	1134
1900	1240	1491	210
1020	360	567	1386
280	1480	1743	1050
1400	600	819	126
520	1720	1995	1302
1640	980	1071	378
760	100	294	1554
1880	1220	1470	630
440	340	546	1806
1560	1460	1722	1029
680		798	105
1800	No. 166	1974	1281
920		462	357
40	**1.** 213724800	1638	1533
1160	**2.** 213920640	714	
420	**3.** 233118720	1890	
1540	**4.** 233314560	966	No. 173
660	**5.** 1039326720	42	
1780	**6.** 330090240	1218	**1.** 224967615
900	**7.** 349676160	441	**2.** 225173757
580	**8.** 1722727680	1617	**3.** 245381736
1700	**9.** 478008960	693	**4.** 245587878
820	**10.** 495672960	1869	**5.** 1093999636
1940	**11.** 427338240	945	**6.** 347454362
1060	**12.** 1039522560	609	**7.** 368070583

8. 1813350334	462	**2.** 236426874	506
9. 503154223	1694	**3.** 257644752	1794
10. 521747423	726	**4.** 257861196	782
11. 449818012	1958	**5.** 1148672552	2070
12. 1094205778	990	**6.** 364818484	1058
13. 388818169	638	**7.** 386465006	46
14. 307038404	1870	**8.** 1903972988	1334
15. 429424101	902	**9.** 528299486	483
	2134	**10.** 547821886	1771
	1166	**11.** 472297784	759
No. 179	198	**12.** 1148888996	2047
	1430	**13.** 408249458	1035
22	616	**14.** 322382728	667
1254	1848	**15.** 450884682	1955
286	880		943
1518	2112		2231
550	1144		1219
1782	792	**No. 186**	207
814	2024		1495
2046	1056	23	644
1232	88	1311	1932
264	1320	299	920
1496	352	1587	2208
528	1584	575	1196
1760	770	1863	828
176	2002	851	2116
1408	1034	2139	1104
440	66	1288	92
1672	1298	276	1380
704	946	1564	368
1936	2178	552	1656
968	1210	1840	805
154	242	184	2093
1386	1474	1472	1081
418	506	460	69
1650	1738	1748	1357
682	924	736	989
1914	2156	2024	2277
330	1188	1012	1265
1562	220	161	253
604	1452	1449	1541
1826	1100	437	529
858	132	1725	1817
2090	1364	713	966
1122	396	2001	2254
308	1628	345	1242
1540	660	1623	230
572	1892	621	1518
1804	1078	1909	1150
836	110	897	138
2068	1342	2185	1426
484	374	1173	414
1716	1606	322	1702
748		1610	690
1980		598	1978
1012	**No. 180**	1886	1127
44		874	115
1276	**1.** 236210430	2162	1403

391	336	1776	1775
1679	1680	720	675
	624	2064	2075
	1968	1176	975
No. 187	912	120	2375
	2256	1464	1275
1. 247453245	528	408	350
2. 247679991	1872	1752	1750
3. 269907768	816		650
4. 270134514	2160		2050
5. 1203345468	1104	**No. 194**	950
6. 382182606	48		2350
7. 404859429	1392	**1.** 258696060	550
8. 1994595642	504	**2.** 258933108	1950
9. 553444749	1848	**3.** 282170784	850
10. 573896349	792	**4.** 282407832	2250
11. 494777556	2136	**5.** 1258018384	1150
12. 1203572214	1080	**6.** 399546728	50
13. 427680747	696	**7.** 423253852	1450
14. 337727052	2040	**8.** 2085218296	525
15. 472345263	984	**9.** 578590012	1925
	2328	**10.** 599970812	825
	1272	**11.** 517257328	2225
No. 193	216	**12.** 1258255432	1125
	1560	**13.** 447112036	725
24	672	**14.** 353071376	2125
1368	2016	**15.** 493805844	1025
312	960		2425
1656	2304		1325
600	1248	**No. 200**	225
1944	864		1625
888	2208	25	700
2232	1152	1425	2100
1344	96	325	1000
288	1440	1725	2400
1632	384	625	1300
576	1728	2025	900
1920	840	925	2300
192	2184	2325	1200
1536	1128	1400	100
480	72	300	1500
1824	1416	1700	400
768	1032	600	1800
2112	2376	2000	875
1056	1320	200	2275
168	264	1600	1175
1512	1608	500	75
456	552	1900	1475
1800	1896	800	1075
744	1008	2200	2475
2088	2352	1100	1375
360	1296	175	275
1704	240	1575	1675
648	1584	475	575
1992	1200	1875	1975
936	144	775	1050
2280	1488	2175	2450
1224	432	375	1350

250
1650
1250
150
1550
450
1850
750
2150
1225
125
1525
425
1825

No. 201

1. 269938875
2. 270186225
3. 294433800
4. 294681150
5. 1312691300
6. 416910850
7. 441648275
3. 2175840950
9. 603735275
10. 626045275
11. 539737100
12. 1312938650
13. 466543325
14. 368415700
15. 515266425

No. 204

(*Annex 0 to
Answers to
No. 45*)

No. 208

(*Annex 0 to
Answers to
No. 46*)

No. 212

(*Annex 0 to
Answers to
No. 47*)

No. 215

(*Annex 0 to
Answers to
No. 50*)

No. 219

(*Annex 0 to
Answers to
No. 52*)

No. 222

(*Annex 0 to
Answers to
No. 53*)

No. 226

(*Annex 0 to
Answers to
No. 56*)

No. 228

(*Annex 0 to
Answers to
No. 60*)

No. 229

1. 242
2. 464
3. 686
4. 902
5. 1124
6. 1246
7. 1462
8. 1684
9. 1906
10. 322
11. 444
12. 666
13. 882
14. 1104
15. 1326
16. 1442
17. 1664
18. 1886
19. 302
20. 524

No. 232

(*Annex 0 to
Answers to
No. 61*)

No. 233

1. 393

2. 726
3. 1059
4. 1392
5. 1713
6. 1896
7. 2229
8. 2562
9. 2883
10. 516
11. 699
12. 1032
13. 1353
14. 1686
15. 2019
16. 2202
17. 2523
18. 2856
19. 489
20. 822

No. 236

(*Annex 0 to
Answers to
No. 77*)

No. 237

1. 564
2. 1008
3. 1452
4. 1896
5. 2340
6. 2564
7. 3008
8. 3452
9. 3892
10. 740
11. 964
12. 1408
13. 1852
14. 2296
15. 2740
16. 2964
17. 3408
18. 3852
19. 696
20. 1140

No. 239

(*Annex 0 to
Answers to
No. 90*)

No. 240

1. 755
2. 1310
3. 1865
4. 2420
5. 2975
6. 3280
7. 3805
8. 4360
9. 4915
10. 970
11. 1275
12. 1830
13. 2355
14. 2910
15. 3465
16. 3770
17. 4325
18. 4880
19. 905
20. 1460

No. 242

(*Annex 0 to
Answers to
No. 106*)

No. 243

1. 846
2. 1512
3. 2178
4. 2844
5. 3510
6. 4176
7. 4482
8. 5106
9. 5772
10. 1038
11. 1704
12. 2370
13. 2676
14. 3342
15. 3966
16. 4632
17. 5298
18. 5964
19. 870
20. 1536

No. 244

(*Annex 0 to
Answers to
No. 119*)

No. 245

1. 917
2. 1694
3. 2471
4. 3248
5. 4025
6. 4802
7. 5579
8. 5866
9. 6587
10. 1064
11. 1841
12. 2618
13. 3395
14. 4172
15. 4459
16. 5236
17. 5957
18. 6734
19. 1211
20. 1988

No. 246

(Annex 0 to
Answers to
No. 131)

No. 247

1. 1128
2. 2016
3. 2904
4. 3792
5. 4680
6. 5568
7. 5976
8. 6864
9. 7752
10. 1368
11. 2256
12. 3144
13. 3552
14. 4440
15. 5328
16. 6216
17. 7104
18. 7992
19. 5928
20. 5216

No. 248

1. $\frac{4}{8}$, $\frac{2}{8}$, $\frac{6}{8}$

2. $\frac{2}{16}$, $\frac{4}{16}$, $\frac{6}{16}$, $\frac{8}{16}$, $\frac{10}{16}$, $\frac{12}{16}$, $\frac{14}{16}$, $\frac{16}{16}$
3. $\frac{2}{8}$, $\frac{4}{8}$, $\frac{6}{8}$
4. $\frac{2}{12}$, $\frac{3}{12}$, $\frac{4}{12}$, $\frac{6}{12}$, $\frac{8}{12}$, $\frac{9}{12}$, $\frac{10}{12}$, $\frac{12}{12}$
5. $\frac{2}{24}$, $\frac{3}{24}$, $\frac{4}{24}$, $\frac{6}{24}$, $\frac{8}{24}$, $\frac{10}{24}$, $\frac{12}{24}$, $\frac{14}{24}$, $\frac{15}{24}$, $\frac{16}{24}$, $\frac{18}{24}$, $\frac{20}{24}$, $\frac{22}{24}$
6. $\frac{2}{10}$, $\frac{4}{10}$, $\frac{5}{10}$, $\frac{6}{10}$, $\frac{8}{10}$
7. $\frac{2}{20}$, $\frac{4}{20}$, $\frac{6}{20}$, $\frac{8}{20}$, $\frac{10}{20}$, $\frac{12}{20}$, $\frac{14}{20}$, $\frac{16}{20}$, $\frac{18}{20}$
8. $\frac{4}{40}$, $\frac{5}{40}$, $\frac{6}{40}$, $\frac{10}{40}$, $\frac{12}{40}$, $\frac{15}{40}$, $\frac{16}{40}$, $\frac{20}{40}$, $\frac{24}{40}$, $\frac{18}{40}$, $\frac{20}{40}$, $\frac{24}{40}$, $\frac{25}{40}$, $\frac{28}{40}$, $\frac{30}{40}$, $\frac{32}{40}$, $\frac{35}{40}$, $\frac{36}{40}$, $\frac{40}{40}$
9. $\frac{5}{15}$, $\frac{6}{15}$, $\frac{8}{15}$, $\frac{9}{15}$, $\frac{10}{15}$, $\frac{12}{15}$
10. $\frac{3}{30}$, $\frac{5}{30}$, $\frac{6}{30}$, $\frac{9}{30}$, $\frac{10}{30}$, $\frac{12}{30}$, $\frac{15}{30}$, $\frac{18}{30}$, $\frac{20}{30}$, $\frac{21}{30}$, $\frac{24}{30}$, $\frac{25}{30}$, $\frac{27}{30}$

No. 249

(Annex 0 to
Answers to
No. 140)

No. 250

1. $\frac{3}{4}$
2. $1\frac{1}{4}$
3. $\frac{5}{8}$
4. $\frac{7}{8}$
5. $1\frac{1}{2}$
6. $1\frac{3}{8}$
7. $\frac{3}{8}$
8. $\frac{5}{8}$
9. $\frac{7}{8}$
10. $1\frac{1}{8}$
11. $\frac{7}{8}$
12. $1\frac{1}{2}$
13. $1\frac{3}{8}$
14. $1\frac{5}{8}$
15. $\frac{9}{16}$
16. $\frac{11}{16}$
17. $\frac{13}{16}$
18. $\frac{15}{16}$

19. $1\frac{1}{16}$
20. $1\frac{3}{16}$
21. $1\frac{5}{16}$
22. $1\frac{7}{16}$
23. $\frac{5}{16}$
24. $\frac{7}{16}$
25. $\frac{9}{16}$
26. $\frac{11}{16}$
27. $\frac{13}{16}$
28. $\frac{15}{16}$
29. $1\frac{1}{16}$
30. $1\frac{3}{16}$
31. $\frac{13}{16}$
32. $\frac{15}{16}$
33. $1\frac{1}{16}$
34. $1\frac{3}{16}$
35. $1\frac{5}{16}$
36. $1\frac{7}{16}$
37. $1\frac{1}{16}$
38. $1\frac{11}{16}$
39. $\frac{3}{16}$
40. $\frac{5}{16}$

No. 251

1. 1368
2. 2367
3. 3366
4. 4365
5. 5364
6. 5823
7. 6822
8. 7821
9. 8757
10. 1656
11. 2655
12. 3114
13. 4113
14. 5112
15. 6111
16. 7056
17. 8055
18. 8514
19. 1413
20. 2412

No. 252

1. 121
2. 232
3. 343
4. 451
5. 562
6. 623
7. 731
8. 842

9. 953
10. 161
11. 222
12. 333
13. 441
14. 552
15. 663
16. 721
17. 832
18. 943
19. 151
20. 262

No. 253

1. $\frac{7}{16}$
2. $\frac{9}{16}$
3. $\frac{11}{16}$
4. $\frac{13}{16}$
5. $\frac{15}{16}$
6. $1\frac{1}{16}$
7. $\frac{7}{16}$
8. $\frac{9}{16}$
9. $\frac{11}{16}$
10. $\frac{13}{16}$

No. 254

(Annex 0 to
Answers to
No. 148)

No. 255

1. 131
2. 242
3. 353
4. 464
5. 571
6. 632
7. 743
8. 854
9. 961
10. 172
11. 233
12. 344
13. 451
14. 562
15. 673
16. 734
17. 841
18. 952
19. 163
20. 274

No. 256
1. $\frac{15}{16}$
2. $1\frac{1}{16}$
3. $1\frac{3}{16}$
4. $1\frac{5}{16}$
5. $\frac{11}{16}$
6. $\frac{13}{16}$
7. $\frac{15}{16}$
8. $1\frac{1}{16}$
9. $1\frac{3}{16}$
10. $1\frac{5}{16}$

No. 257
(*Annex 0 to Answers to No. 156*)

No. 258
1. 141
2. 252
3. 363
4. 474
5. 585
6. 641
7. 752
8. 863
9. 974
10. 185
11. 241
12. 352
13. 463
14. 574
15. 685
16. 741
17. 852
18. 963
19. 174
20. 285

No. 259
1. $1\frac{7}{16}$
2. $1\frac{9}{16}$
3. $\frac{15}{16}$
4. $1\frac{1}{16}$
5. $1\frac{3}{16}$
6. $1\frac{5}{16}$
7. $1\frac{7}{16}$
8. $1\frac{9}{16}$
9. $1\frac{11}{16}$
10. $1\frac{13}{16}$

No. 260
(*Annex 0 to Answers to No. 165*)

No. 261
1. $\frac{1}{2}$
2. $\frac{5}{6}$
3. $\frac{5}{12}$
4. $\frac{3}{4}$
5. $1\frac{11}{12}$
6. $1\frac{1}{4}$
7. $\frac{3}{4}$
8. $1\frac{1}{12}$
9. $1\frac{1}{4}$
10. $1\frac{7}{12}$

No. 262
1. 151
2. 262
3. 373
4. 484
5. 595
6. 656
7. 761
8. 872
9. 983
10. 194
11. 255
12. 366
13. 471
14. 582
15. 693
16. 754
17. 865
18. 976
19. 181
20. 292

No. 263
1. $\frac{1}{4}$
2. $\frac{3}{4}$
3. $\frac{1}{8}$
4. $\frac{3}{8}$
5. $\frac{5}{8}$
6. $\frac{7}{8}$
7. $\frac{1}{8}$
8. $\frac{3}{8}$
9. $\frac{5}{8}$
10. $\frac{7}{8}$
11. $\frac{1}{8}$
12. $\frac{3}{8}$
13. $\frac{5}{8}$
14. $\frac{7}{8}$
15. $\frac{1}{16}$
16. $\frac{3}{16}$
17. $\frac{5}{16}$
18. $\frac{7}{16}$
19. $\frac{9}{16}$
20. $\frac{11}{16}$
21. $\frac{13}{16}$
22. $\frac{15}{16}$
23. $\frac{1}{16}$
24. $\frac{3}{16}$
25. $\frac{5}{16}$
26. $\frac{7}{16}$
27. $\frac{9}{16}$
28. $\frac{11}{16}$
29. $\frac{13}{16}$
30. $\frac{15}{16}$

No. 264
(*Annex 0 to Answers to No. 172*)

No. 265
1. $\frac{1}{16}$
2. $\frac{3}{16}$
3. $\frac{5}{16}$
4. $\frac{7}{16}$
5. $\frac{9}{16}$
6. $\frac{11}{16}$
7. $\frac{13}{16}$
8. $\frac{15}{16}$
9. $\frac{1}{16}$
10. $\frac{3}{16}$

No. 266
1. 141
2. 252
3. 363
4. 474
5. 585
6. 696
7. 747
8. 851
9. 962
10. 173
11. 284
12. 395
13. 446
14. 557
15. 661
16. 772
17. 883
18. 994
19. 145
20. 256

No. 267
1. $\frac{1}{4}$
2. $\frac{7}{12}$
3. $\frac{3}{4}$
4. $1\frac{1}{12}$
5. $1\frac{11}{12}$
6. $1\frac{1}{4}$
7. $1\frac{5}{12}$
8. $1\frac{1}{4}$
9. $\frac{5}{6}$
10. $1\frac{1}{6}$

No. 268
(*Annex 0 to Answers to No. 179*)

No. 269
1. $\frac{5}{16}$
2. $\frac{7}{16}$
3. $\frac{9}{16}$
4. $\frac{11}{16}$
5. $\frac{13}{16}$
6. $\frac{15}{16}$
7. $\frac{1}{16}$
8. $\frac{3}{16}$
9. $\frac{5}{16}$
10. $\frac{7}{16}$

No. 270
1. 131
2. 242
3. 353
4. 464
5. 575
6. 686
7. 797
8. 838
9. 941
10. 152
11. 263
12. 374
13. 485

14. 596
15. 637
16. 748
17. 851
18. 962
19. 173
20. 284

No. 271

1. $\frac{2}{3}$
2. $1\frac{1}{3}$
3. $\frac{5}{12}$
4. $1\frac{1}{12}$
5. $1\frac{1}{12}$
6. $1\frac{7}{12}$
7. $\frac{7}{24}$
8. $\frac{11}{24}$
9. $\frac{19}{24}$
10. $1\frac{19}{24}$

No. 272

(*Annex 0 to Answers to No. 186*)

No. 273

1. $\frac{9}{16}$
2. $\frac{11}{16}$
3. $\frac{13}{16}$
4. $\frac{15}{16}$
5. $\frac{1}{16}$
6. $\frac{3}{16}$
7. $\frac{5}{16}$
8. $\frac{7}{16}$
9. $\frac{9}{16}$
10. $\frac{11}{16}$

No. 274

1. 141
2. 252
3. 363
4. 474
5. 585
6. 696
7. 747
8. 858
9. 969
10. 171
11. 282

12. 393
13. 444
14. 555
15. 666
16. 777
17. 888
18. 999
19. 741
20. 652

No. 275

1. $\frac{23}{24}$
2. $1\frac{5}{24}$
3. $1\frac{11}{24}$
4. $1\frac{17}{24}$
5. $\frac{7}{12}$
6. $\frac{11}{12}$
7. $1\frac{1}{12}$
8. $1\frac{5}{12}$
9. $\frac{1}{3}$
10. $\frac{2}{3}$

No. 276

(*Annex 0 to Answers to No. 193*)

No. 277

1. $\frac{13}{16}$
2. $\frac{15}{16}$
3. $\frac{1}{16}$
4. $\frac{3}{16}$
5. $\frac{5}{16}$
6. $\frac{7}{16}$
7. $\frac{9}{16}$
8. $\frac{11}{16}$
9. $\frac{13}{16}$
10. $\frac{15}{16}$

No. 278

1. 152
2. 263
3. 374
4. 485
5. 596
6. 647
7. 758

8. 869
9. 973
10. 184
11. 295
12. 346
13. 437
14. 568
15. 679
16. 784
17. 895
18. 946
19. 157
20. 268

No. 279

1. $\frac{5}{6}$
2. $1\frac{1}{6}$
3. $\frac{5}{6}$
4. $1\frac{1}{6}$
5. $1\frac{1}{6}$
6. $1\frac{2}{3}$
7. $\frac{5}{24}$
8. $\frac{13}{24}$
9. $\frac{17}{24}$
10. $1\frac{1}{24}$

No. 280

(*Annex 0 to Answers to No. 200*)

No. 281

1. $\frac{1}{6}$
2. $\frac{1}{6}$
3. $\frac{1}{12}$
4. $\frac{5}{12}$
5. $\frac{7}{12}$
6. $1\frac{1}{12}$
7. $\frac{1}{12}$
8. $\frac{5}{12}$
9. $\frac{7}{12}$
10. $1\frac{1}{12}$

No. 282

1. 2r86
2. 2r129
3. 2r108
4. 2r347
5. 2r456
6. 2r589

7. 2r312
8. 2r102
9. 2r208
10. 2r117
11. 3r13
12. 3r50
13. 3r105
14. 3r182
15. 3r285
16. 4r126
17. 4r200
18. 4r252
19. 4r282
20. 4r280

No. 283

1. $\frac{11}{24}$
2. $\frac{19}{24}$
3. $\frac{23}{24}$
4. $1\frac{7}{24}$
5. $\frac{11}{24}$
6. $1\frac{5}{24}$
7. $1\frac{5}{24}$
8. $1\frac{13}{24}$
9. $\frac{23}{24}$
10. $1\frac{7}{24}$

No. 284

1. 1066
2. 1377
3. 1708
4. 2059
5. 2511
6. 2912
7. 1023
8. 1394
9. 1326
10. 1647
11. 1988
12. 2349
13. 2821
14. 992
15. 1353
16. 1734
17. 1586
18. 1917
19. 2268
20. 2639

No. 285

1. $\frac{1}{12}$
2. $\frac{5}{12}$
3. $\frac{7}{12}$

4. $\frac{11}{12}$
5. $\frac{1}{12}$
6. $\frac{5}{12}$
7. $\frac{7}{12}$
8. $\frac{11}{12}$
9. $\frac{1}{2}$
10. $\frac{2}{3}$

No. 286

1. 2r1
2. 2r29
3. 2r376
4. 2r551
5. 2r374
6. 3r378
7. 3r518
8. 3r680
9. 3r864
10. 3r17
11. 4r266
12. 4r225
13. 4r172
14. 4r93
15. 4r162
16. 5r90
17. 5r130
18. 5r148
19. 5r144
20. 5r119

No. 287

1. $1\frac{11}{24}$
2. $1\frac{19}{24}$
3. $\frac{3}{16}$
4. $\frac{1}{2}$
5. $\frac{9}{10}$
6. $1\frac{1}{10}$
7. $\frac{1}{2}$
8. $\frac{7}{10}$
9. $1\frac{1}{10}$
10. $1\frac{3}{10}$

No. 288

1. 1470
2. 1872
3. 2294
4. 2736
5. 3198
6. 3772
7. 1344

8. 1806
9. 1820
10. 2232
11. 2664
12. 3116
13. 3588
14. 1312
15. 1764
16. 2236
17. 2108
18. 2520
19. 2952
20. 3404

No. 289

1. $\frac{1}{6}$
2. $\frac{5}{6}$
3. $\frac{5}{6}$
4. $\frac{19}{24}$
5. $\frac{1}{6}$
6. $\frac{5}{6}$
7. $\frac{5}{6}$
8. $\frac{5}{6}$
9. $\frac{1}{6}$
10. $\frac{1}{6}$

No. 290

1. 2r37
2. 2r771
3. 2r150
4. 2r85
5. 2r99
6. 3r46
7. 3r102
8. 3r170
9. 3r280
10. 3r402
11. 4r192
12. 4r235
13. 4r276
14. 4r285
15. 4r272
16. 5r67
17. 5r693
18. 5r564
19. 5r632
20. 5r97

No. 291

1. $\frac{7}{10}$
2. $\frac{7}{10}$

3. $1\frac{3}{10}$
4. $1\frac{1}{2}$
5. $\frac{9}{10}$
6. $1\frac{1}{10}$
7. $1\frac{1}{2}$
8. $1\frac{7}{10}$
9. $\frac{7}{10}$
10. $\frac{9}{10}$

No. 292

1. 1892
2. 2385
3. 2898
4. 3431
5. 3984
6. 4557
7. 1683
8. 2236
9. 2332
10. 2835
11. 3358
12. 3901
13. 4464
14. 1617
15. 2193
16. 2756
17. 2772
18. 3510
19. 3818
20. 4371

No. 293

1. $\frac{5}{6}$
2. $\frac{5}{6}$
3. $\frac{5}{6}$
4. $\frac{5}{6}$
5. $\frac{1}{12}$
6. $\frac{5}{12}$
7. $\frac{7}{12}$
8. $\frac{11}{12}$
9. $\frac{1}{12}$
10. $\frac{5}{12}$

No. 294

1. 3r51
2. 3r69
3. 3r95
4. 3r32
5. 3r54
6. 4r226
7. 4r85
8. 4r864

9. 4r119
10. 4r208
11. 5r146
12. 5r288
13. 5r321
14. 5r465
15. 5r108
16. 6r125
17. 6r200
18. 6r77
19. 6r111
20. 6r310

No. 295

1. $1\frac{1}{10}$
2. $1\frac{3}{10}$
3. $\frac{3}{5}$
4. $\frac{4}{5}$
5. $1\frac{1}{5}$
6. $1\frac{2}{5}$
7. $\frac{9}{20}$
8. $\frac{13}{20}$
9. $\frac{17}{20}$
10. $1\frac{1}{20}$

No. 296

1. 2332
2. 2916
3. 3520
4. 4144
5. 4788
6. 5452
7. 2006
8. 2684
9. 2862
10. 3456
11. 4070
12. 4704
14. 1972
15. 2596
16. 3599
17. 3392
18. 3996
19. 4620
20. 5264

No. 297

1. $\frac{7}{12}$
2. $1\frac{1}{12}$
3. $\frac{1}{12}$
4. $\frac{5}{12}$

5. $\frac{7}{12}$
6. $\frac{11}{12}$
7. $\frac{1}{12}$
8. $\frac{5}{12}$
9. $\frac{7}{12}$
10. $\frac{11}{12}$

No. 298
1. 5r219
2. 5r642
3. 5r312
4. 5r97
5. 5r106
6. 6r310
7. 6r150
8. 6r100
9. 6r609
10. 6r115
11. 7r65
12. 7r135
13. 7r235
14. 7r185
15. 7r64
16. 8r72
17. 8r125
18. 8r180
19. 8r360
20. 8r421

No. 299
1. $\frac{7}{20}$
2. $\frac{11}{20}$
3. $\frac{19}{20}$
4. $1\frac{8}{20}$
5. $\frac{19}{20}$
6. $1\frac{8}{20}$
7. $1\frac{7}{20}$
8. $1\frac{11}{20}$
9. $1\frac{7}{20}$
10. $1\frac{1}{20}$

No. 300
1. 2790
2. 3465
3. 4160
4. 4875
5. 5610
6. 6365
7. 2380
8. 3105

9. 3410
10. 4095
11. 4800
12. 5525
13. 6270
14. 2345
15. 3060
16. 3795
17. 4030
18. 4725
19. 5440
20. 6175

No. 301
1. $\frac{1}{12}$
2. $\frac{5}{12}$
3. $\frac{7}{12}$
4. $\frac{11}{12}$
5. $\frac{1}{12}$
6. $\frac{5}{12}$
7. $\frac{7}{12}$
8. $\frac{11}{12}$
9. $\frac{1}{12}$
10. $\frac{5}{12}$

No. 302
1. 6r10
2. 6r29
3. 6r38
4. 6r165
5. 6r651
6. 7r501
7. 7r307
8. 7r799
9. 7r646
10. 7r20
11. 8r189
12. 8r612
13. 8r325
14. 8r486
15. 8r17
16. 9r125
17. 9r135
18. 9r74
19. 9r85
20. 9r59

No. 303
1. $1\frac{9}{20}$
2. $1\frac{18}{20}$
3. $\frac{18}{20}$
4. $\frac{21}{40}$

5. $\frac{21}{40}$
6. $\frac{27}{40}$
7. $\frac{9}{40}$
8. $\frac{17}{40}$
9. $\frac{33}{40}$
10. $1\frac{1}{40}$

No. 304
1. 3266
2. 4032
3. 4818
4. 5624
5. 6450
6. 7296
7. 2772
8. 3588
9. 3976
10. 4752
11. 5548
12. 6364
13. 7200
14. 2736
15. 3542
16. 4368
17. 4686
18. 5472
19. 6278
20. 7104

No. 305
1. $\frac{7}{12}$
2. $\frac{11}{12}$
3. $\frac{1}{10}$
4. $\frac{8}{10}$
5. $\frac{7}{10}$
6. $\frac{9}{10}$
7. $\frac{1}{10}$
8. $\frac{8}{10}$
9. $\frac{7}{10}$
10. $\frac{9}{10}$

No. 306
1. 6r706
2. 6r95
3. 6r37
4. 6r38
5. 6r40
6. 7r18
7. 7r118
8. 7r211
9. 7r346
10. 7r252
11. 8r28
12. 8r39

13. 8r404
14. 8r355
15. 8r626
16. 9r64
17. 9r301
18. 9r400
19. 9r500
20. 9r65

No. 307
1. $\frac{23}{40}$
2. $\frac{31}{40}$
3. $\frac{39}{40}$
4. $1\frac{7}{40}$
5. $\frac{13}{40}$
6. $\frac{27}{40}$
7. $1\frac{3}{40}$
8. $1\frac{11}{40}$
9. $\frac{33}{40}$
10. $1\frac{1}{40}$

No. 308
1. 3713
2. 4617
3. 5494
4. 6391
5. 7308
6. 8245
7. 3182
8. 4089
9. 4503
10. 5427
11. 6314
12. 7221
13. 8148
14. 3145
15. 4042
16. 4959
17. 5293
18. 6237
19. 7134
20. 8051

No. 309
1. $\frac{1}{10}$
2. $\frac{1}{10}$
3. $\frac{7}{10}$
4. $\frac{1}{10}$
5. $\frac{1}{10}$
6. $\frac{8}{10}$
7. $\frac{7}{10}$
8. $\frac{9}{10}$

9. $\frac{1}{8}$
10. $\frac{3}{8}$

No. 310

1. 7r129
2. 7r642
3. 7r711
4. 7r32
5. 7r232
6. 8r77
7. 8r444
8. 8r312
9. 8r147
10. 8r25
11. 9r27
12. 9r297
13. 9r358
14. 9r555
15. 9r609
16. 9r775
17. 9r862
18. 9r927
19. 9r150
20. 9r215

No. 311

1. $1\frac{9}{40}$
2. $1\frac{17}{40}$
3. $\frac{29}{40}$
4. $\frac{37}{40}$
5. $1\frac{13}{40}$
6. $1\frac{21}{40}$
7. $1\frac{3}{40}$
8. $1\frac{11}{40}$
9. $1\frac{19}{40}$
10. $1\frac{27}{40}$

No. 312

1. 4224
2. 5162
3. 6188
4. 7176
5. 8184
6. 9212
7. 3610
8. 4608
9. 5104
10. 6052
11. 7098
12. 8096
13. 9114
14. 3572

15. 4560
16. 5568
17. 5984
18. 6942
19. 8008
20. 9016

No. 313

1. $\frac{3}{5}$
2. $\frac{4}{5}$
3. $\frac{7}{10}$
4. $\frac{8}{10}$
5. $\frac{7}{10}$
6. $\frac{9}{10}$
7. $\frac{1}{5}$
8. $\frac{2}{5}$
9. $\frac{3}{5}$
10. $\frac{4}{5}$

No. 314

1. $\frac{38}{40}$
2. $1\frac{7}{40}$
3. $1\frac{28}{40}$
4. $1\frac{31}{40}$
5. $\frac{8}{15}$
6. $\frac{11}{15}$
7. $\frac{14}{15}$
8. $1\frac{2}{15}$
9. $1\frac{18}{30}$
10. $1\frac{8}{30}$

No. 315

1. 4655
2. 5664
3. 6693
4. 7742
5. 8811
6. 9405
7. 3744
8. 4753
9. 5782
10. 6831
11. 7505
12. 8544
13. 9603
14. 3822
15. 4851
16. 5605
17. 6624
18. 7663
19. 8722
20. 9801

No. 316

1. $\frac{1}{10}$
2. $\frac{8}{10}$
3. $\frac{7}{10}$
4. $\frac{9}{10}$
5. $\frac{1}{5}$
6. $\frac{2}{5}$
7. $\frac{3}{5}$
8. $\frac{4}{5}$
9. $\frac{1}{10}$
10. $\frac{3}{10}$

No. 317

1. $1\frac{1}{30}$
2. $1\frac{7}{30}$
3. $1\frac{12}{15}$
4. $1\frac{1}{15}$
5. $1\frac{4}{15}$
6. $1\frac{7}{15}$
7. $1\frac{28}{30}$
8. $1\frac{29}{30}$
9. $1\frac{1}{30}$
10. $1\frac{17}{30}$

No. 318

1. $\frac{7}{10}$
2. $\frac{9}{10}$
3. $\frac{1}{10}$
4. $\frac{2}{5}$
5. $\frac{3}{5}$
6. $\frac{4}{5}$
7. $\frac{1}{10}$
8. $\frac{3}{10}$
9. $\frac{7}{10}$
10. $\frac{9}{10}$

No. 319

1. 41
2. 51
3. 61
4. 71
5. 81
6. 91
7. 31
8. 41
9. 51
10. 61
11. 71
12. 81
13. 91

14. 31
15. 41
16. 51
17. 61
18. 71
19. 81
20. 91

No. 320

1. $\frac{11}{30}$
2. $\frac{17}{30}$
3. $\frac{23}{30}$
4. $\frac{29}{30}$
5. $\frac{4}{15}$
6. $\frac{7}{15}$
7. $\frac{13}{15}$
8. $1\frac{1}{15}$
9. $1\frac{1}{30}$
10. $1\frac{7}{30}$

No. 321

1. $\frac{1}{5}$
2. $\frac{3}{5}$
3. $\frac{3}{5}$
4. $\frac{4}{5}$
5. $\frac{1}{10}$
6. $\frac{3}{10}$
7. $\frac{7}{10}$
8. $\frac{9}{10}$
9. $\frac{3}{5}$
10. $\frac{3}{5}$

No. 322

1. 42
2. 52
3. 62
4. 72
5. 82
6. 92
7. 32
8. 42
9. 52
10. 62
11. 72
12. 82
13. 92
14. 32
15. 42
16. 52
17. 62
18. 72
19. 82

20. 92

No. 323

1. $1\frac{13}{30}$
2. $1\frac{19}{30}$
3. $1\frac{1}{4}$
4. $1\frac{2}{15}$
5. $1\frac{8}{15}$
6. $1\frac{11}{15}$

No. 324

1. $\frac{3}{8}$
2. $\frac{4}{5}$
3. $\frac{1}{10}$
4. $\frac{3}{10}$
5. $\frac{7}{10}$
6. $\frac{9}{10}$
7. $\frac{1}{4}$
8. $\frac{3}{8}$
9. $\frac{3}{8}$
10. $\frac{3}{4}$

No. 325

1. 43
2. 53
3. 63
4. 73
5. 83
6. 93
7. 33
8. 43
9. 53
10. 63
11. 73
12. 83
13. 93
14. 33
15. 43
16. 53
17. 63
18. 73
19. 83
20. 93

No. 327

1. $\frac{1}{10}$
2. $\frac{3}{10}$
3. $\frac{7}{10}$
4. $\frac{9}{10}$
5. $\frac{1}{5}$

6. $\frac{2}{5}$
7. $\frac{3}{5}$
8. $\frac{4}{5}$
9. $\frac{1}{10}$
10. $\frac{3}{10}$

No. 328

1. 44
2. 54
3. 64
4. 74
5. 84
6. 94
7. 34
8. 44
9. 54
10. 64
11. 74
12. 84
13. 94
14. 34
15. 44
16. 54
17. 64
18. 74
19. 84
20. 94

No. 330

1. $\frac{7}{10}$
2. $\frac{9}{10}$
3. $\frac{1}{5}$
4. $\frac{2}{5}$
5. $\frac{3}{5}$
6. $\frac{4}{5}$
7. $\frac{1}{10}$
8. $\frac{3}{10}$
9. $\frac{7}{10}$
10. $\frac{9}{10}$

No. 331

1. 45
2. 55
3. 65
4. 75
5. 85
6. 95
7. 35
8. 45
9. 55
10. 65
11. 75
12. 85

13. 95
14. 35
15. 45
16. 55
17. 65
18. 75
19. 85
20. 95

No. 332

1. 46
2. 56
3. 66
4. 76
5. 86
6. 96
7. 36
8. 46
9. 56
10. 66
11. 76
12. 86
13. 96
14. 36
15. 46
16. 56
17. 66
18. 76
19. 86
20. 96

No. 333

1. $\frac{1}{5}$
2. $\frac{2}{5}$
3. $\frac{3}{5}$
4. $\frac{4}{5}$
5. $\frac{1}{10}$
6. $\frac{3}{10}$
7. $\frac{7}{10}$
8. $\frac{9}{10}$
9. $\frac{1}{5}$
10. $\frac{2}{5}$

No. 334

1. 47
2. 57
3. 67
4. 77
5. 87
6. 97
7. 37
8. 47
9. 57

10. 67
11. 77
12. 87
13. 97
14. 37
15. 47
16. 57
17. 67
18. 77
19. 87
20. 97

No. 335

1. $\frac{3}{8}$
2. $\frac{4}{5}$
3. $\frac{1}{10}$
4. $\frac{3}{10}$
5. $\frac{7}{10}$
6. $\frac{9}{10}$

No. 336

1. 48
2. 58
3. 68
4. 78
5. 88
6. 98
7. 38
8. 48
9. 58
10. 68
11. 78
12. 88
13. 98
14. 38
15. 48
16. 58
17. 68
18. 78
19. 88
20. 98

No. 337

1. 49
2. 59
3. 69
4. 79
5. 89
6. 99
7. 39
8. 49
9. 59
10. 69

11. 79
12. 89
13. 99
14. 39
15. 49
16. 59
17. 69
18. 79
19. 89
20. 99

No. 338

1. .12½
2. .37½
3. .62½
4. .87½
5. .33⅓
6. .66⅔
7. .16⅔
8. .83⅓
9. .20
10. .40
11. .60
12. .80

No. 339

1. 2886
2. 5994
3. 9268
4. 12818
5. 17081
6. 19584
7. 23793
8. 28288
9. 24466
10. 4104

No. 340

1. .06¼
2. .18¾
3. .31¼
4. .43¾
5. .56¼
6. .68¾
7. .81¼
8. .93¾
9. .08⅓
10. .41⅔
11. .58⅓
12. .91⅔
13. .03⅛
14. .04⅙

No. 341

1. 4235
2. 8352
3. 12691
4. 17138
5. 21918
6. 25543
7. 30702
8. 36206
9. 33355
10. 5796

No. 342

1. $17887
2. $9818
3. 9865
4. 25775
5. 39540
6. 23332
7. 17313
8. 31383
9. $14822.40
10. 243062

No. 343

1. 5764
2. 10890
3. 16238
4. 21808
5. 27408
6. 30968
7. 37893
8. 44408
9. 42284
10. 7740

No. 344

1. .0625
2. .1875
3. .3125
4. .4375
5. .5625
6. .6875
7. .8125
8. .9375
9. .0833⅓
10. .4166⅔
11. .5833⅓
12. .9166⅔
13. .0312½
14. .0416⅔

No. 345

1. 7473
2. 13608
3. 19965
4. 26544
5. 33345
6. 37178
7. 44368
8. 52643
9. 51622
10. 9990

No. 346

1. $99.84
2. 96256
3. $117.76
4. 98304
5. 1728
6. $675.84
7. $8120.60
8. $30402.55

No. 347

1. 9362
2. 16506
3. 23872
4. 31460
5. 39270
6. 43952
7. 51748
8. 60168
9. 60946
10. 12222

No. 348

1. .03125
2. .09375
3. .15625
4. .21875
5. .28125
6. .34375
7. .40625
8. .46875
9. .53125
10. .59375
11. .65625
12. .71875
13. .78125
14. .84375
15. .90625
16. .96875
17. .04167

18. .20833
19. .29167
20. .45833
21. .54167
22. .70833
23. .79167
24. .95833

No. 349

1. 10011
2. 18144
3. 26499
4. 35076
5. 43875
6. 52896
7. 57519
8. 66378
9. 68302
10. 12456

No. 350

1. $424575
2. $84770
3. $733779.50
4. $26863.20
5. $830062.74
6. $526.32
7. $981088
8. $9603
9. $1007010

No. 351

1. 10349
2. 19602
3. 28946
4. 38512
5. 48300
6. 58310
7. 68542
8. 72906
9. 74339
10. 12312

No. 353

1. 12408
2. 22428
3. 33033
4. 43608
5. 54405
6. 65424

7. 70965
8. 82368
9. 85272
10. 15219

No. 354

1. $525
2. $756
3. $384
4. $810
5. $5400
6. $900
7. $13000
8. $14700
9. $7200
10. $1600
11. $630
12. $12600
13. $1200
14. $1200
15. $1200

No. 355

1. 14440
2. 25248
3. 36278
4. 47530
5. 59004
6. 61465
7. 72768
8. 84293
9. 95354
10. 19206

No. 357

1. 11211
2. 24642
3. 40051
4. 57902
5. 77691
6. 92412
7. 116081
8. 142272
9. 170321
10. 29032

No. 358

1. $247715.70
2. $243540
3. $60226335
4. $1087638.75

5. $5209451.52
6. $131602.24
7. $40102686.72
8. $8710669

No. 359

1. 24442
2. 49184
3. 76146
4. 104632
5. 136004
6. 156996
7. 191522
8. 229024
9. 268746
10. 47012

No. 361

1. 39693
2. 75746
3. 114019
4. 154512
5. 195853
6. 223096
7. 269709
8. 318542
9. 368063
10. 67596

No. 362

1. 138138
2. 115596
3. 74556
4. 186960
5. 89301
6. 235872
7. 119782
8. 73248
9. 193256

No. 363

1. 56964
2. 104328
3. 153912
4. 205716
5. 259740
6. 291014
7. 348928
8. 409062
9. 471416
10. 91390

No. 364

1. 210
2. 342
3. 255
4. 240
5. 195
6. 247
7. 272
8. 224
9. 361

No. 365

1. 76255
2. 134930
3. 195825
4. 258940
5. 324275
6. 364080
7. 429965
8. 501400
9. 575055
10. 115430

No. 366

1. $56496
2. $799018
3. $5663152
4. $410091.55
5. $453952.95
6. $36033.25
7. $530895.75
8. $1043606.30

No. 367

1.   85446
2. 155232
3. 227238
4. 301464
5. 377910
6. 456576
7. 497502
8. 575276
9. 659932
10. 120408

No. 368

1. $139510.50

2. $147804.75
3. $158233.30
4. $131011.65
5. $452339.40
6. $754503.75
7. $151524.65
8. $238939.80

No. 369

1. 92617
2. 173514
3. 256631
4. 341968
5. 429525
6. 519302
7. 611299
8. 651126
9. 740567
10. 121144

No. 370

1. 5476
2. 8649
3. 6724
4. 4096
5. 1444
6. 12544
7. 15376
8. 21316
9. 28224
10. 38809
11. 1236544
12. 1471369
13. 1726596
14. 2298256
15. 2954961

No. 371

1. 113928
2. 206136
3. 300564
4. 397212
5. 496080
6. 597168
7. 648396
8. 753324
9. 860472
10. 153558

No. 372

1. 7616

2. 12561
3. 15824
4. 22425
5. 40716
6. 42749
7. 421056
8. 224196
9. 198989

No. 373

1. 138168
2. 241697
3. 347446
4. 455415
5. 565604
6. 620473
7. 734502
8. 850751
9. 962297
10. 183816

No. 374

1. 8556
2. 4030
3. 7308
4. 8924
5. 45795
6. 100152
7. 173888
8. 264171
9. 837221

No. 375

1. 2025
2. 3025

3. 4225
4. 5625
5. 7225
6. 9025
7. 13225
8. 18225
9. 24025
10. 30625
11. 38025
12. 99225
13. 112225
14. 126025
15. 140625

No. 376

1. 621
2. 2009
3. 1224
4. 11021
5. 13216
6. 24024
7. 30616
8. 27209
9. 38016

No. 377

1. 275625
2. 390625
3. 680625
4. 1050625
5. 1500625
6. 1755625
7. 2640625
8. 2975625

9. 3330625
10. 3705625

No. 378

1. 4896
2. 6391
3. 8084
4. 12019
5. 16851
6. 22484
7. 25536
8. 32351
9. 36036

No. 379

1. $90\frac{2}{3}$
2. $112\frac{9}{25}$
3. $160\frac{5}{12}$
4. $339\frac{1}{6}$
5. $12\frac{4}{5}$
6. $3681\frac{9}{20}$
7. $1625\frac{3}{32}$
8. $650\frac{2}{5}$
9. $28\frac{7}{8}$
10. $72\frac{11}{12}$
11. $42\frac{15}{64}$
12. $152\frac{5}{81}$

No. 380

1. 276
2. 800
3. $929\frac{1}{3}$
4. 950

5. 2552
6. 5952
7. 1422
8. 2100
9. 3363

No. 381

1. 23.2
2. 45
3. 36
4. 3.5
5. 5.12
6. 13.05
7. 10.18
8. 61.2
9. 77.6

No. 382

1. 2744
2. 19683
3. 35937
4. 97336
5. 205379
6. 238328
7. 274625
8. 357911
9. 389017
10. 592704
11. 636056
12. 681472
13. 857375
14. 912673
15. 970299

A CATALOG OF SELECTED
DOVER BOOKS
IN ALL FIELDS OF INTEREST

A CATALOG OF SELECTED DOVER
BOOKS IN ALL FIELDS OF INTEREST

CONCERNING THE SPIRITUAL IN ART, Wassily Kandinsky. Pioneering work by father of abstract art. Thoughts on color theory, nature of art. Analysis of earlier masters. 12 illustrations. 80pp. of text. 5⅜ x 8½. 0-486-23411-8

CELTIC ART: The Methods of Construction, George Bain. Simple geometric techniques for making Celtic interlacements, spirals, Kells-type initials, animals, humans, etc. Over 500 illustrations. 160pp. 9 x 12. (Available in U.S. only.) 0-486-22923-8

AN ATLAS OF ANATOMY FOR ARTISTS, Fritz Schider. Most thorough reference work on art anatomy in the world. Hundreds of illustrations, including selections from works by Vesalius, Leonardo, Goya, Ingres, Michelangelo, others. 593 illustrations. 192pp. 7⅛ x 10¼. 0-486-20241-0

CELTIC HAND STROKE-BY-STROKE (Irish Half-Uncial from "The Book of Kells"): An Arthur Baker Calligraphy Manual, Arthur Baker. Complete guide to creating each letter of the alphabet in distinctive Celtic manner. Covers hand position, strokes, pens, inks, paper, more. Illustrated. 48pp. 8¼ x 11. 0-486-24336-2

EASY ORIGAMI, John Montroll. Charming collection of 32 projects (hat, cup, pelican, piano, swan, many more) specially designed for the novice origami hobbyist. Clearly illustrated easy-to-follow instructions insure that even beginning papercrafters will achieve successful results. 48pp. 8¼ x 11. 0-486-27298-2

BLOOMINGDALE'S ILLUSTRATED 1886 CATALOG: Fashions, Dry Goods and Housewares, Bloomingdale Brothers. Famed merchants' extremely rare catalog depicting about 1,700 products: clothing, housewares, firearms, dry goods, jewelry, more. Invaluable for dating, identifying vintage items. Also, copyright-free graphics for artists, designers. Co-published with Henry Ford Museum & Greenfield Village. 160pp. 8¼ x 11. 0-486-25780-0

THE ART OF WORLDLY WISDOM, Baltasar Gracian. "Think with the few and speak with the many," "Friends are a second existence," and "Be able to forget" are among this 1637 volume's 300 pithy maxims. A perfect source of mental and spiritual refreshment, it can be opened at random and appreciated either in brief or at length. 128pp. 5⅜ x 8½. 0-486-44034-6

JOHNSON'S DICTIONARY: A Modern Selection, Samuel Johnson (E. L. McAdam and George Milne, eds.). This modern version reduces the original 1755 edition's 2,300 pages of definitions and literary examples to a more manageable length, retaining the verbal pleasure and historical curiosity of the original. 480pp. 5³⁄₁₆ x 8¼. 0-486-44089-3

ADVENTURES OF HUCKLEBERRY FINN, Mark Twain, Illustrated by E. W. Kemble. A work of eternal richness and complexity, a source of ongoing critical debate, and a literary landmark, Twain's 1885 masterpiece about a barefoot boy's journey of self-discovery has enthralled readers around the world. This handsome clothbound reproduction of the first edition features all 174 of the original black-and-white illustrations. 368pp. 5⅜ x 8½. 0-486-44322-1

CATALOG OF DOVER BOOKS

STICKLEY CRAFTSMAN FURNITURE CATALOGS, Gustav Stickley and L. & J. G. Stickley. Beautiful, functional furniture in two authentic catalogs from 1910. 594 illustrations, including 277 photos, show settles, rockers, armchairs, reclining chairs, bookcases, desks, tables. 183pp. 6½ x 9¼. 0-486-23838-5

AMERICAN LOCOMOTIVES IN HISTORIC PHOTOGRAPHS: 1858 to 1949, Ron Ziel (ed.). A rare collection of 126 meticulously detailed official photographs, called "builder portraits," of American locomotives that majestically chronicle the rise of steam locomotive power in America. Introduction. Detailed captions. xi+ 129pp. 9 x 12. 0-486-27393-8

AMERICA'S LIGHTHOUSES: An Illustrated History, Francis Ross Holland, Jr. Delightfully written, profusely illustrated fact-filled survey of over 200 American lighthouses since 1716. History, anecdotes, technological advances, more. 240pp. 8 x 10¾. 0-486-25576-X

TOWARDS A NEW ARCHITECTURE, Le Corbusier. Pioneering manifesto by founder of "International School." Technical and aesthetic theories, views of industry, economics, relation of form to function, "mass-production split" and much more. Profusely illustrated. 320pp. 6⅛ x 9¼. (Available in U.S. only.) 0-486-25023-7

HOW THE OTHER HALF LIVES, Jacob Riis. Famous journalistic record, exposing poverty and degradation of New York slums around 1900, by major social reformer. 100 striking and influential photographs. 233pp. 10 x 7⅞. 0-486-22012-5

FRUIT KEY AND TWIG KEY TO TREES AND SHRUBS, William M. Harlow. One of the handiest and most widely used identification aids. Fruit key covers 120 deciduous and evergreen species; twig key 160 deciduous species. Easily used. Over 300 photographs. 126pp. 5⅜ x 8½. 0-486-20511-8

COMMON BIRD SONGS, Dr. Donald J. Borror. Songs of 60 most common U.S. birds: robins, sparrows, cardinals, bluejays, finches, more–arranged in order of increasing complexity. Up to 9 variations of songs of each species.
Cassette and manual 0-486-99911-4

ORCHIDS AS HOUSE PLANTS, Rebecca Tyson Northen. Grow cattleyas and many other kinds of orchids–in a window, in a case, or under artificial light. 63 illustrations. 148pp. 5⅜ x 8½. 0-486-23261-1

MONSTER MAZES, Dave Phillips. Masterful mazes at four levels of difficulty. Avoid deadly perils and evil creatures to find magical treasures. Solutions for all 32 exciting illustrated puzzles. 48pp. 8¼ x 11. 0-486-26005-4

MOZART'S DON GIOVANNI (DOVER OPERA LIBRETTO SERIES), Wolfgang Amadeus Mozart. Introduced and translated by Ellen H. Bleiler. Standard Italian libretto, with complete English translation. Convenient and thoroughly portable–an ideal companion for reading along with a recording or the performance itself. Introduction. List of characters. Plot summary. 121pp. 5¼ x 8½. 0-486-24944-1

FRANK LLOYD WRIGHT'S DANA HOUSE, Donald Hoffmann. Pictorial essay of residential masterpiece with over 160 interior and exterior photos, plans, elevations, sketches and studies. 128pp. 9¼ x 10¾. 0-486-29120-0

THE CLARINET AND CLARINET PLAYING, David Pino. Lively, comprehensive work features suggestions about technique, musicianship, and musical interpretation, as well as guidelines for teaching, making your own reeds, and preparing for public performance. Includes an intriguing look at clarinet history. "A godsend," *The Clarinet,* Journal of the International Clarinet Society. Appendixes. 7 illus. 320pp. 5⅜ x 8½.   0-486-40270-3

HOLLYWOOD GLAMOR PORTRAITS, John Kobal (ed.). 145 photos from 1926-49. Harlow, Gable, Bogart, Bacall; 94 stars in all. Full background on photographers, technical aspects. 160pp. 8⅜ x 11¼.   0-486-23352-9

THE RAVEN AND OTHER FAVORITE POEMS, Edgar Allan Poe. Over 40 of the author's most memorable poems: "The Bells," "Ulalume," "Israfel," "To Helen," "The Conqueror Worm," "Eldorado," "Annabel Lee," many more. Alphabetic lists of titles and first lines. 64pp. 5³⁄₁₆ x 8¼.   0-486-26685-0

PERSONAL MEMOIRS OF U. S. GRANT, Ulysses Simpson Grant. Intelligent, deeply moving firsthand account of Civil War campaigns, considered by many the finest military memoirs ever written. Includes letters, historic photographs, maps and more. 528pp. 6⅛ x 9¼.   0-486-28587-1

ANCIENT EGYPTIAN MATERIALS AND INDUSTRIES, A. Lucas and J. Harris. Fascinating, comprehensive, thoroughly documented text describes this ancient civilization's vast resources and the processes that incorporated them in daily life, including the use of animal products, building materials, cosmetics, perfumes and incense, fibers, glazed ware, glass and its manufacture, materials used in the mummification process, and much more. 544pp. 6⅛ x 9¼. (Available in U.S. only.)
0-486-40446-3

RUSSIAN STORIES/RUSSKIE RASSKAZY: A Dual-Language Book, edited by Gleb Struve. Twelve tales by such masters as Chekhov, Tolstoy, Dostoevsky, Pushkin, others. Excellent word-for-word English translations on facing pages, plus teaching and study aids, Russian/English vocabulary, biographical/critical introductions, more. 416pp. 5⅜ x 8½.   0-486-26244-8

PHILADELPHIA THEN AND NOW: 60 Sites Photographed in the Past and Present, Kenneth Finkel and Susan Oyama. Rare photographs of City Hall, Logan Square, Independence Hall, Betsy Ross House, other landmarks juxtaposed with contemporary views. Captures changing face of historic city. Introduction. Captions. 128pp. 8¼ x 11.   0-486-25790-8

NORTH AMERICAN INDIAN LIFE: Customs and Traditions of 23 Tribes, Elsie Clews Parsons (ed.). 27 fictionalized essays by noted anthropologists examine religion, customs, government, additional facets of life among the Winnebago, Crow, Zuni, Eskimo, other tribes. 480pp. 6⅛ x 9¼.   0-486-27377-6

TECHNICAL MANUAL AND DICTIONARY OF CLASSICAL BALLET, Gail Grant. Defines, explains, comments on steps, movements, poses and concepts. 15-page pictorial section. Basic book for student, viewer. 127pp. 5⅜ x 8½.
0-486-21843-0

THE MALE AND FEMALE FIGURE IN MOTION: 60 Classic Photographic Sequences, Eadweard Muybridge. 60 true-action photographs of men and women walking, running, climbing, bending, turning, etc., reproduced from rare 19th-century masterpiece. vi + 121pp. 9 x 12.   0-486-24745-7

ANIMALS: 1,419 Copyright-Free Illustrations of Mammals, Birds, Fish, Insects, etc., Jim Harter (ed.). Clear wood engravings present, in extremely lifelike poses, over 1,000 species of animals. One of the most extensive pictorial sourcebooks of its kind. Captions. Index. 284pp. 9 x 12. 0-486-23766-4

1001 QUESTIONS ANSWERED ABOUT THE SEASHORE, N. J. Berrill and Jacquelyn Berrill. Queries answered about dolphins, sea snails, sponges, starfish, fishes, shore birds, many others. Covers appearance, breeding, growth, feeding, much more. 305pp. 5¼ x 8¼. 0-486-23366-9

ATTRACTING BIRDS TO YOUR YARD, William J. Weber. Easy-to-follow guide offers advice on how to attract the greatest diversity of birds: birdhouses, feeders, water and waterers, much more. 96pp. 5³⁄₁₆ x 8¼. 0-486-28927-3

MEDICINAL AND OTHER USES OF NORTH AMERICAN PLANTS: A Historical Survey with Special Reference to the Eastern Indian Tribes, Charlotte Erichsen-Brown. Chronological historical citations document 500 years of usage of plants, trees, shrubs native to eastern Canada, northeastern U.S. Also complete identifying information. 343 illustrations. 544pp. 6½ x 9¼. 0-486-25951-X

STORYBOOK MAZES, Dave Phillips. 23 stories and mazes on two-page spreads: Wizard of Oz, Treasure Island, Robin Hood, etc. Solutions. 64pp. 8¼ x 11.
0-486-23628-5

AMERICAN NEGRO SONGS: 230 Folk Songs and Spirituals, Religious and Secular, John W. Work. This authoritative study traces the African influences of songs sung and played by black Americans at work, in church, and as entertainment. The author discusses the lyric significance of such songs as "Swing Low, Sweet Chariot," "John Henry," and others and offers the words and music for 230 songs. Bibliography. Index of Song Titles. 272pp. 6½ x 9¼. 0-486-40271-1

MOVIE-STAR PORTRAITS OF THE FORTIES, John Kobal (ed.). 163 glamor, studio photos of 106 stars of the 1940s: Rita Hayworth, Ava Gardner, Marlon Brando, Clark Gable, many more. 176pp. 8⅜ x 11¼. 0-486-23546-7

YEKL and THE IMPORTED BRIDEGROOM AND OTHER STORIES OF YIDDISH NEW YORK, Abraham Cahan. Film Hester Street based on *Yekl* (1896). Novel, other stories among first about Jewish immigrants on N.Y.'s East Side. 240pp. 5⅜ x 8½. 0-486-22427-9

SELECTED POEMS, Walt Whitman. Generous sampling from *Leaves of Grass*. Twenty-four poems include "I Hear America Singing," "Song of the Open Road," "I Sing the Body Electric," "When Lilacs Last in the Dooryard Bloom'd," "O Captain! My Captain!"—all reprinted from an authoritative edition. Lists of titles and first lines. 128pp. 5³⁄₁₆ x 8¼. 0-486-26878-0

SONGS OF EXPERIENCE: Facsimile Reproduction with 26 Plates in Full Color, William Blake. 26 full-color plates from a rare 1826 edition. Includes "The Tyger," "London," "Holy Thursday," and other poems. Printed text of poems. 48pp. 5¼ x 7.
0-486-24636-1

THE BEST TALES OF HOFFMANN, E. T. A. Hoffmann. 10 of Hoffmann's most important stories: "Nutcracker and the King of Mice," "The Golden Flowerpot," etc. 458pp. 5⅜ x 8½. 0-486-21793-0

THE BOOK OF TEA, Kakuzo Okakura. Minor classic of the Orient: entertaining, charming explanation, interpretation of traditional Japanese culture in terms of tea ceremony. 94pp. 5⅜ x 8½. 0-486-20070-1

FRENCH STORIES/CONTES FRANÇAIS: A Dual-Language Book, Wallace Fowlie. Ten stories by French masters, Voltaire to Camus: "Micromegas" by Voltaire; "The Atheist's Mass" by Balzac; "Minuet" by de Maupassant; "The Guest" by Camus, six more. Excellent English translations on facing pages. Also French-English vocabulary list, exercises, more. 352pp. 5⅜ x 8½. 0-486-26443-2

CHICAGO AT THE TURN OF THE CENTURY IN PHOTOGRAPHS: 122 Historic Views from the Collections of the Chicago Historical Society, Larry A. Viskochil. Rare large-format prints offer detailed views of City Hall, State Street, the Loop, Hull House, Union Station, many other landmarks, circa 1904-1913. Introduction. Captions. Maps. 144pp. 9⅜ x 12¼. 0-486-24656-6

OLD BROOKLYN IN EARLY PHOTOGRAPHS, 1865-1929, William Lee Younger. Luna Park, Gravesend race track, construction of Grand Army Plaza, moving of Hotel Brighton, etc. 157 previously unpublished photographs. 165pp. 8⅞ x 11¾. 0-486-23587-4

THE MYTHS OF THE NORTH AMERICAN INDIANS, Lewis Spence. Rich anthology of the myths and legends of the Algonquins, Iroquois, Pawnees and Sioux, prefaced by an extensive historical and ethnological commentary. 36 illustrations. 480pp. 5⅜ x 8½. 0-486-25967-6

AN ENCYCLOPEDIA OF BATTLES: Accounts of Over 1,560 Battles from 1479 B.C. to the Present, David Eggenberger. Essential details of every major battle in recorded history from the first battle of Megiddo in 1479 B.C. to Grenada in 1984. List of Battle Maps. New Appendix covering the years 1967-1984. Index. 99 illustrations. 544pp. 6½ x 9¼. 0-486-24913-1

SAILING ALONE AROUND THE WORLD, Captain Joshua Slocum. First man to sail around the world, alone, in small boat. One of great feats of seamanship told in delightful manner. 67 illustrations. 294pp. 5⅜ x 8½. 0-486-20326-3

ANARCHISM AND OTHER ESSAYS, Emma Goldman. Powerful, penetrating, prophetic essays on direct action, role of minorities, prison reform, puritan hypocrisy, violence, etc. 271pp. 5⅜ x 8½. 0-486-22484-8

MYTHS OF THE HINDUS AND BUDDHISTS, Ananda K. Coomaraswamy and Sister Nivedita. Great stories of the epics; deeds of Krishna, Shiva, taken from puranas, Vedas, folk tales; etc. 32 illustrations. 400pp. 5⅜ x 8½. 0-486-21759-0

MY BONDAGE AND MY FREEDOM, Frederick Douglass. Born a slave, Douglass became outspoken force in antislavery movement. The best of Douglass' autobiographies. Graphic description of slave life. 464pp. 5⅜ x 8½. 0-486-22457-0

FOLLOWING THE EQUATOR: A Journey Around the World, Mark Twain. Fascinating humorous account of 1897 voyage to Hawaii, Australia, India, New Zealand, etc. Ironic, bemused reports on peoples, customs, climate, flora and fauna, politics, much more. 197 illustrations. 720pp. 5⅜ x 8½. 0-486-26113-1

THE PEOPLE CALLED SHAKERS, Edward D. Andrews. Definitive study of Shakers: origins, beliefs, practices, dances, social organization, furniture and crafts, etc. 33 illustrations. 351pp. 5⅜ x 8½. 0-486-21081-2

THE MYTHS OF GREECE AND ROME, H. A. Guerber. A classic of mythology, generously illustrated, long prized for its simple, graphic, accurate retelling of the principal myths of Greece and Rome, and for its commentary on their origins and significance. With 64 illustrations by Michelangelo, Raphael, Titian, Rubens, Canova, Bernini and others. 480pp. 5⅜ x 8½. 0-486-27584-1

PSYCHOLOGY OF MUSIC, Carl E. Seashore. Classic work discusses music as a medium from psychological viewpoint. Clear treatment of physical acoustics, auditory apparatus, sound perception, development of musical skills, nature of musical feeling, host of other topics. 88 figures. 408pp. 5⅜ x 8½. 0-486-21851-1

LIFE IN ANCIENT EGYPT, Adolf Erman. Fullest, most thorough, detailed older account with much not in more recent books, domestic life, religion, magic, medicine, commerce, much more. Many illustrations reproduce tomb paintings, carvings, hieroglyphs, etc. 597pp. 5⅜ x 8½. 0-486-22632-8

SUNDIALS, Their Theory and Construction, Albert Waugh. Far and away the best, most thorough coverage of ideas, mathematics concerned, types, construction, adjusting anywhere. Simple, nontechnical treatment allows even children to build several of these dials. Over 100 illustrations. 230pp. 5⅜ x 8½. 0-486-22947-5

THEORETICAL HYDRODYNAMICS, L. M. Milne-Thomson. Classic exposition of the mathematical theory of fluid motion, applicable to both hydrodynamics and aerodynamics. Over 600 exercises. 768pp. 6⅛ x 9¼. 0-486-68970-0

OLD-TIME VIGNETTES IN FULL COLOR, Carol Belanger Grafton (ed.). Over 390 charming, often sentimental illustrations, selected from archives of Victorian graphics–pretty women posing, children playing, food, flowers, kittens and puppies, smiling cherubs, birds and butterflies, much more. All copyright-free. 48pp. 9¼ x 12¼. 0-486-27269-9

PERSPECTIVE FOR ARTISTS, Rex Vicat Cole. Depth, perspective of sky and sea, shadows, much more, not usually covered. 391 diagrams, 81 reproductions of drawings and paintings. 279pp. 5⅜ x 8½. 0-486-22487-2

DRAWING THE LIVING FIGURE, Joseph Sheppard. Innovative approach to artistic anatomy focuses on specifics of surface anatomy, rather than muscles and bones. Over 170 drawings of live models in front, back and side views, and in widely varying poses. Accompanying diagrams. 177 illustrations. Introduction. Index. 144pp. 8⅜ x11¼. 0-486-26723-7

GOTHIC AND OLD ENGLISH ALPHABETS: 100 Complete Fonts, Dan X. Solo. Add power, elegance to posters, signs, other graphics with 100 stunning copyright-free alphabets: Blackstone, Dolbey, Germania, 97 more–including many lower-case, numerals, punctuation marks. 104pp. 8⅛ x 11. 0-486-24695-7

THE BOOK OF WOOD CARVING, Charles Marshall Sayers. Finest book for beginners discusses fundamentals and offers 34 designs. "Absolutely first rate . . . well thought out and well executed."–E. J. Tangerman. 118pp. 7¾ x 10⅝. 0-486-23654-4

ILLUSTRATED CATALOG OF CIVIL WAR MILITARY GOODS: Union Army Weapons, Insignia, Uniform Accessories, and Other Equipment, Schuyler, Hartley, and Graham. Rare, profusely illustrated 1846 catalog includes Union Army uniform and dress regulations, arms and ammunition, coats, insignia, flags, swords, rifles, etc. 226 illustrations. 160pp. 9 x 12. 0-486-24939-5

WOMEN'S FASHIONS OF THE EARLY 1900s: An Unabridged Republication of "New York Fashions, 1909," National Cloak & Suit Co. Rare catalog of mail-order fashions documents women's and children's clothing styles shortly after the turn of the century. Captions offer full descriptions, prices. Invaluable resource for fashion, costume historians. Approximately 725 illustrations. 128pp. 8⅜ x 11¼. 0-486-27276-1

HOW TO DO BEADWORK, Mary White. Fundamental book on craft from simple projects to five-bead chains and woven works. 106 illustrations. 142pp. 5⅜ x 8. 0-486-20697-1

THE 1912 AND 1915 GUSTAV STICKLEY FURNITURE CATALOGS, Gustav Stickley. With over 200 detailed illustrations and descriptions, these two catalogs are essential reading and reference materials and identification guides for Stickley furniture. Captions cite materials, dimensions and prices. 112pp. 6½ x 9¼. 0-486-26676-1

EARLY AMERICAN LOCOMOTIVES, John H. White, Jr. Finest locomotive engravings from early 19th century: historical (1804–74), main-line (after 1870), special, foreign, etc. 147 plates. 142pp. 11⅜ x 8¼. 0-486-22772-3

LITTLE BOOK OF EARLY AMERICAN CRAFTS AND TRADES, Peter Stockham (ed.). 1807 children's book explains crafts and trades: baker, hatter, cooper, potter, and many others. 23 copperplate illustrations. 140pp. 4⅝ x 6. 0-486-23336-7

VICTORIAN FASHIONS AND COSTUMES FROM HARPER'S BAZAR, 1867–1898, Stella Blum (ed.). Day costumes, evening wear, sports clothes, shoes, hats, other accessories in over 1,000 detailed engravings. 320pp. 9⅜ x 12¼. 0-486-22990-4

THE LONG ISLAND RAIL ROAD IN EARLY PHOTOGRAPHS, Ron Ziel. Over 220 rare photos, informative text document origin (1844) and development of rail service on Long Island. Vintage views of early trains, locomotives, stations, passengers, crews, much more. Captions. 8⅞ x 11¾. 0-486-26301-0

VOYAGE OF THE LIBERDADE, Joshua Slocum. Great 19th-century mariner's thrilling, first-hand account of the wreck of his ship off South America, the 35-foot boat he built from the wreckage, and its remarkable voyage home. 128pp. 5⅜ x 8½. 0-486-40022-0

TEN BOOKS ON ARCHITECTURE, Vitruvius. The most important book ever written on architecture. Early Roman aesthetics, technology, classical orders, site selection, all other aspects. Morgan translation. 331pp. 5⅜ x 8½. 0-486-20645-9

THE HUMAN FIGURE IN MOTION, Eadweard Muybridge. More than 4,500 stopped-action photos, in action series, showing undraped men, women, children jumping, lying down, throwing, sitting, wrestling, carrying, etc. 390pp. 7⅞ x 10⅝. 0-486-20204-6 Clothbd.

TREES OF THE EASTERN AND CENTRAL UNITED STATES AND CANADA, William M. Harlow. Best one-volume guide to 140 trees. Full descriptions, woodlore, range, etc. Over 600 illustrations. Handy size. 288pp. 4½ x 6⅜. 0-486-20395-6

GROWING AND USING HERBS AND SPICES, Milo Miloradovich. Versatile handbook provides all the information needed for cultivation and use of all the herbs and spices available in North America. 4 illustrations. Index. Glossary. 236pp. 5⅜ x 8½. 0-486-25058-X

BIG BOOK OF MAZES AND LABYRINTHS, Walter Shepherd. 50 mazes and labyrinths in all–classical, solid, ripple, and more–in one great volume. Perfect inexpensive puzzler for clever youngsters. Full solutions. 112pp. 8⅛ x 11. 0-486-22951-3

PIANO TUNING, J. Cree Fischer. Clearest, best book for beginner, amateur. Simple repairs, raising dropped notes, tuning by easy method of flattened fifths. No previous skills needed. 4 illustrations. 201pp. 5⅜ x 8½. 0-486-23267-0

HINTS TO SINGERS, Lillian Nordica. Selecting the right teacher, developing confidence, overcoming stage fright, and many other important skills receive thoughtful discussion in this indispensible guide, written by a world-famous diva of four decades' experience. 96pp. 5⅜ x 8½. 0-486-40094-8

THE COMPLETE NONSENSE OF EDWARD LEAR, Edward Lear. All nonsense limericks, zany alphabets, Owl and Pussycat, songs, nonsense botany, etc., illustrated by Lear. Total of 320pp. 5⅜ x 8½. (Available in U.S. only.) 0-486-20167-8

VICTORIAN PARLOUR POETRY: An Annotated Anthology, Michael R. Turner. 117 gems by Longfellow, Tennyson, Browning, many lesser-known poets. "The Village Blacksmith," "Curfew Must Not Ring Tonight," "Only a Baby Small," dozens more, often difficult to find elsewhere. Index of poets, titles, first lines. xxiii + 325pp. 5⅝ x 8¼. 0-486-27044-0

DUBLINERS, James Joyce. Fifteen stories offer vivid, tightly focused observations of the lives of Dublin's poorer classes. At least one, "The Dead," is considered a masterpiece. Reprinted complete and unabridged from standard edition. 160pp. $5\frac{3}{16}$ x 8¼. 0-486-26870-5

GREAT WEIRD TALES: 14 Stories by Lovecraft, Blackwood, Machen and Others, S. T. Joshi (ed.). 14 spellbinding tales, including "The Sin Eater," by Fiona McLeod, "The Eye Above the Mantel," by Frank Belknap Long, as well as renowned works by R. H. Barlow, Lord Dunsany, Arthur Machen, W. C. Morrow and eight other masters of the genre. 256pp. 5⅜ x 8½. (Available in U.S. only.) 0-486-40436-6

THE BOOK OF THE SACRED MAGIC OF ABRAMELIN THE MAGE, translated by S. MacGregor Mathers. Medieval manuscript of ceremonial magic. Basic document in Aleister Crowley, Golden Dawn groups. 268pp. 5⅜ x 8½. 0-486-23211-5

THE BATTLES THAT CHANGED HISTORY, Fletcher Pratt. Eminent historian profiles 16 crucial conflicts, ancient to modern, that changed the course of civilization. 352pp. 5⅜ x 8½. 0-486-41129-X

NEW RUSSIAN-ENGLISH AND ENGLISH-RUSSIAN DICTIONARY, M. A. O'Brien. This is a remarkably handy Russian dictionary, containing a surprising amount of information, including over 70,000 entries. 366pp. 4½ x 6⅛. 0-486-20208-9

NEW YORK IN THE FORTIES, Andreas Feininger. 162 brilliant photographs by the well-known photographer, formerly with *Life* magazine. Commuters, shoppers, Times Square at night, much else from city at its peak. Captions by John von Hartz. 181pp. 9¼ x 10¾. 0-486-23585-8

INDIAN SIGN LANGUAGE, William Tomkins. Over 525 signs developed by Sioux and other tribes. Written instructions and diagrams. Also 290 pictographs. 111pp. 6⅛ x 9¼. 0-486-22029-X

ANATOMY: A Complete Guide for Artists, Joseph Sheppard. A master of figure drawing shows artists how to render human anatomy convincingly. Over 460 illustrations. 224pp. 8⅜ x 11¼. 0-486-27279-6

MEDIEVAL CALLIGRAPHY: Its History and Technique, Marc Drogin. Spirited history, comprehensive instruction manual covers 13 styles (ca. 4th century through 15th). Excellent photographs; directions for duplicating medieval techniques with modern tools. 224pp. 8⅜ x 11¼. 0-486-26142-5

DRIED FLOWERS: How to Prepare Them, Sarah Whitlock and Martha Rankin. Complete instructions on how to use silica gel, meal and borax, perlite aggregate, sand and borax, glycerine and water to create attractive permanent flower arrangements. 12 illustrations. 32pp. 5⅜ x 8½. 0-486-21802-3

EASY-TO-MAKE BIRD FEEDERS FOR WOODWORKERS, Scott D. Campbell. Detailed, simple-to-use guide for designing, constructing, caring for and using feeders. Text, illustrations for 12 classic and contemporary designs. 96pp. 5⅜ x 8½. 0-486-25847-5

THE COMPLETE BOOK OF BIRDHOUSE CONSTRUCTION FOR WOOD-WORKERS, Scott D. Campbell. Detailed instructions, illustrations, tables. Also data on bird habitat and instinct patterns. Bibliography. 3 tables. 63 illustrations in 15 figures. 48pp. 5¼ x 8½. 0-486-24407-5

SCOTTISH WONDER TALES FROM MYTH AND LEGEND, Donald A. Mackenzie. 16 lively tales tell of giants rumbling down mountainsides, of a magic wand that turns stone pillars into warriors, of gods and goddesses, evil hags, powerful forces and more. 240pp. 5⅜ x 8½. 0-486-29677-6

THE HISTORY OF UNDERCLOTHES, C. Willett Cunnington and Phyllis Cunnington. Fascinating, well-documented survey covering six centuries of English undergarments, enhanced with over 100 illustrations: 12th-century laced-up bodice, footed long drawers (1795), 19th-century bustles, l9th-century corsets for men, Victorian "bust improvers," much more. 272pp. 5⅜ x 8¼. 0-486-27124-2

ARTS AND CRAFTS FURNITURE: The Complete Brooks Catalog of 1912, Brooks Manufacturing Co. Photos and detailed descriptions of more than 150 now very collectible furniture designs from the Arts and Crafts movement depict davenports, settees, buffets, desks, tables, chairs, bedsteads, dressers and more, all built of solid, quarter-sawed oak. Invaluable for students and enthusiasts of antiques, Americana and the decorative arts. 80pp. 6½ x 9¼. 0-486-27471-3

WILBUR AND ORVILLE: A Biography of the Wright Brothers, Fred Howard. Definitive, crisply written study tells the full story of the brothers' lives and work. A vividly written biography, unparalleled in scope and color, that also captures the spirit of an extraordinary era. 560pp. 6⅛ x 9¼. 0-486-40297-5

THE ARTS OF THE SAILOR: Knotting, Splicing and Ropework, Hervey Garrett Smith. Indispensable shipboard reference covers tools, basic knots and useful hitches; handsewing and canvas work, more. Over 100 illustrations. Delightful reading for sea lovers. 256pp. 5⅜ x 8½. 0-486-26440-8

FRANK LLOYD WRIGHT'S FALLINGWATER: The House and Its History, Second, Revised Edition, Donald Hoffmann. A total revision—both in text and illustrations—of the standard document on Fallingwater, the boldest, most personal architectural statement of Wright's mature years, updated with valuable new material from the recently opened Frank Lloyd Wright Archives. "Fascinating"–*The New York Times*. 116 illustrations. 128pp. 9¼ x 10¾. 0-486-27430-6

PHOTOGRAPHIC SKETCHBOOK OF THE CIVIL WAR, Alexander Gardner. 100 photos taken on field during the Civil War. Famous shots of Manassas Harper's Ferry, Lincoln, Richmond, slave pens, etc. 244pp. 10⅝ x 8¼. 0-486-22731-6

FIVE ACRES AND INDEPENDENCE, Maurice G. Kains. Great back-to-the-land classic explains basics of self-sufficient farming. The one book to get. 95 illustrations. 397pp. 5⅜ x 8½. 0-486-20974-1

A MODERN HERBAL, Margaret Grieve. Much the fullest, most exact, most useful compilation of herbal material. Gigantic alphabetical encyclopedia, from aconite to zedoary, gives botanical information, medical properties, folklore, economic uses, much else. Indispensable to serious reader. 161 illustrations. 888pp. 6½ x 9¼. 2-vol. set. (Available in U.S. only.)   Vol. I: 0-486-22798-7   Vol. II: 0-486-22799-5

HIDDEN TREASURE MAZE BOOK, Dave Phillips. Solve 34 challenging mazes accompanied by heroic tales of adventure. Evil dragons, people-eating plants, bloodthirsty giants, many more dangerous adversaries lurk at every twist and turn. 34 mazes, stories, solutions. 48pp. 8¼ x 11.   0-486-24566-7

LETTERS OF W. A. MOZART, Wolfgang A. Mozart. Remarkable letters show bawdy wit, humor, imagination, musical insights, contemporary musical world; includes some letters from Leopold Mozart. 276pp. 5⅜ x 8½.   0-486-22859-2

BASIC PRINCIPLES OF CLASSICAL BALLET, Agrippina Vaganova. Great Russian theoretician, teacher explains methods for teaching classical ballet. 118 illustrations. 175pp. 5⅜ x 8½.   0-486-22036-2

THE JUMPING FROG, Mark Twain. Revenge edition. The original story of The Celebrated Jumping Frog of Calaveras County, a hapless French translation, and Twain's hilarious "retranslation" from the French. 12 illustrations. 66pp. 5⅜ x 8½.
0-486-22686-7

BEST REMEMBERED POEMS, Martin Gardner (ed.). The 126 poems in this superb collection of 19th- and 20th-century British and American verse range from Shelley's "To a Skylark" to the impassioned "Renascence" of Edna St. Vincent Millay and to Edward Lear's whimsical "The Owl and the Pussycat." 224pp. 5⅜ x 8½.
0-486-27165-X

COMPLETE SONNETS, William Shakespeare. Over 150 exquisite poems deal with love, friendship, the tyranny of time, beauty's evanescence, death and other themes in language of remarkable power, precision and beauty. Glossary of archaic terms. 80pp. 5³⁄₁₆ x 8¼.   0-486-26686-9

HISTORIC HOMES OF THE AMERICAN PRESIDENTS, Second, Revised Edition, Irvin Haas. A traveler's guide to American Presidential homes, most open to the public, depicting and describing homes occupied by every American President from George Washington to George Bush. With visiting hours, admission charges, travel routes. 175 photographs. Index. 160pp. 8¼ x 11.   0-486-26751-2

THE WIT AND HUMOR OF OSCAR WILDE, Alvin Redman (ed.). More than 1,000 ripostes, paradoxes, wisecracks: Work is the curse of the drinking classes; I can resist everything except temptation; etc. 258pp. 5⅜ x 8½.   0-486-20602-5

SHAKESPEARE LEXICON AND QUOTATION DICTIONARY, Alexander Schmidt. Full definitions, locations, shades of meaning in every word in plays and poems. More than 50,000 exact quotations. 1,485pp. 6½ x 9¼. 2-vol. set.
Vol. 1: 0-486-22726-X   Vol. 2: 0-486-22727-8

SELECTED POEMS, Emily Dickinson. Over 100 best-known, best-loved poems by one of America's foremost poets, reprinted from authoritative early editions. No comparable edition at this price. Index of first lines. 64pp. 5³⁄₁₆ x 8¼. 0-486-26466-1

THE INSIDIOUS DR. FU-MANCHU, Sax Rohmer. The first of the popular mystery series introduces a pair of English detectives to their archnemesis, the diabolical Dr. Fu-Manchu. Flavorful atmosphere, fast-paced action, and colorful characters enliven this classic of the genre. 208pp. 5³⁄₁₆ x 8¼.   0-486-29898-1

THE MALLEUS MALEFICARUM OF KRAMER AND SPRENGER, translated by Montague Summers. Full text of most important witchhunter's "bible," used by both Catholics and Protestants. 278pp. 6⅝ x 10. 0-486-22802-9

SPANISH STORIES/CUENTOS ESPAÑOLES: A Dual-Language Book, Angel Flores (ed.). Unique format offers 13 great stories in Spanish by Cervantes, Borges, others. Faithful English translations on facing pages. 352pp. 5⅜ x 8½. 0-486-25399-6

GARDEN CITY, LONG ISLAND, IN EARLY PHOTOGRAPHS, 1869–1919, Mildred H. Smith. Handsome treasury of 118 vintage pictures, accompanied by carefully researched captions, document the Garden City Hotel fire (1899), the Vanderbilt Cup Race (1908), the first airmail flight departing from the Nassau Boulevard Aerodrome (1911), and much more. 96pp. 8⅞ x 11¾. 0-486-40669-5

OLD QUEENS, N.Y., IN EARLY PHOTOGRAPHS, Vincent F. Seyfried and William Asadorian. Over 160 rare photographs of Maspeth, Jamaica, Jackson Heights, and other areas. Vintage views of DeWitt Clinton mansion, 1939 World's Fair and more. Captions. 192pp. 8⅞ x 11. 0-486-26358-4

CAPTURED BY THE INDIANS: 15 Firsthand Accounts, 1750-1870, Frederick Drimmer. Astounding true historical accounts of grisly torture, bloody conflicts, relentless pursuits, miraculous escapes and more, by people who lived to tell the tale. 384pp. 5⅜ x 8½. 0-486-24901-8

THE WORLD'S GREAT SPEECHES (Fourth Enlarged Edition), Lewis Copeland, Lawrence W. Lamm, and Stephen J. McKenna. Nearly 300 speeches provide public speakers with a wealth of updated quotes and inspiration–from Pericles' funeral oration and William Jennings Bryan's "Cross of Gold Speech" to Malcolm X's powerful words on the Black Revolution and Earl of Spenser's tribute to his sister, Diana, Princess of Wales. 944pp. 5⅜ x 8⅜. 0-486-40903-1

THE BOOK OF THE SWORD, Sir Richard F. Burton. Great Victorian scholar/adventurer's eloquent, erudite history of the "queen of weapons"–from prehistory to early Roman Empire. Evolution and development of early swords, variations (sabre, broadsword, cutlass, scimitar, etc.), much more. 336pp. 6⅛ x 9¼. 0-486-25434-8

AUTOBIOGRAPHY: The Story of My Experiments with Truth, Mohandas K. Gandhi. Boyhood, legal studies, purification, the growth of the Satyagraha (nonviolent protest) movement. Critical, inspiring work of the man responsible for the freedom of India. 480pp. 5⅜ x 8½. (Available in U.S. only.) 0-486-24593-4

CELTIC MYTHS AND LEGENDS, T. W. Rolleston. Masterful retelling of Irish and Welsh stories and tales. Cuchulain, King Arthur, Deirdre, the Grail, many more. First paperback edition. 58 full-page illustrations. 512pp. 5⅜ x 8½. 0-486-26507-2

THE PRINCIPLES OF PSYCHOLOGY, William James. Famous long course complete, unabridged. Stream of thought, time perception, memory, experimental methods; great work decades ahead of its time. 94 figures. 1,391pp. 5⅜ x 8½. 2-vol. set. Vol. I: 0-486-20381-6 Vol. II: 0-486-20382-4

THE WORLD AS WILL AND REPRESENTATION, Arthur Schopenhauer. Definitive English translation of Schopenhauer's life work, correcting more than 1,000 errors, omissions in earlier translations. Translated by E. F. J. Payne. Total of 1,269pp. 5⅜ x 8½. 2-vol. set. Vol. 1: 0-486-21761-2 Vol. 2: 0-486-21762-0

MAGIC AND MYSTERY IN TIBET, Madame Alexandra David-Neel. Experiences among lamas, magicians, sages, sorcerers, Bonpa wizards. A true psychic discovery. 32 illustrations. 321pp. 5⅜ x 8½. (Available in U.S. only.) 0-486-22682-4

THE EGYPTIAN BOOK OF THE DEAD, E. A. Wallis Budge. Complete reproduction of Ani's papyrus, finest ever found. Full hieroglyphic text, interlinear transliteration, word-for-word translation, smooth translation. 533pp. 6½ x 9¼.
0-486-21866-X

HISTORIC COSTUME IN PICTURES, Braun & Schneider. Over 1,450 costumed figures in clearly detailed engravings–from dawn of civilization to end of 19th century. Captions. Many folk costumes. 256pp. 8⅜ x 11¾. 0-486-23150-X

MATHEMATICS FOR THE NONMATHEMATICIAN, Morris Kline. Detailed, college-level treatment of mathematics in cultural and historical context, with numerous exercises. Recommended Reading Lists. Tables. Numerous figures. 641pp. 5⅜ x 8½.
0-486-24823-2

PROBABILISTIC METHODS IN THE THEORY OF STRUCTURES, Isaac Elishakoff. Well-written introduction covers the elements of the theory of probability from two or more random variables, the reliability of such multivariable structures, the theory of random function, Monte Carlo methods of treating problems incapable of exact solution, and more. Examples. 502pp. 5⅜ x 8½. 0-486-40691-1

THE RIME OF THE ANCIENT MARINER, Gustave Doré, S. T. Coleridge. Doré's finest work; 34 plates capture moods, subtleties of poem. Flawless full-size reproductions printed on facing pages with authoritative text of poem. "Beautiful. Simply beautiful."–*Publisher's Weekly.* 77pp. 9¼ x 12. 0-486-22305-1

SCULPTURE: Principles and Practice, Louis Slobodkin. Step-by-step approach to clay, plaster, metals, stone; classical and modern. 253 drawings, photos. 255pp. 8⅛ x 11.
0-486-22960-2

THE INFLUENCE OF SEA POWER UPON HISTORY, 1660–1783, A. T. Mahan. Influential classic of naval history and tactics still used as text in war colleges. First paperback edition. 4 maps. 24 battle plans. 640pp. 5⅜ x 8½. 0-486-25509-3

THE STORY OF THE TITANIC AS TOLD BY ITS SURVIVORS, Jack Winocour (ed.). What it was really like. Panic, despair, shocking inefficiency, and a little heroism. More thrilling than any fictional account. 26 illustrations. 320pp. 5⅜ x 8½.
0-486-20610-6

ONE TWO THREE . . . INFINITY: Facts and Speculations of Science, George Gamow. Great physicist's fascinating, readable overview of contemporary science: number theory, relativity, fourth dimension, entropy, genes, atomic structure, much more. 128 illustrations. Index. 352pp. 5⅜ x 8½. 0-486-25664-2

DALÍ ON MODERN ART: The Cuckolds of Antiquated Modern Art, Salvador Dalí. Influential painter skewers modern art and its practitioners. Outrageous evaluations of Picasso, Cézanne, Turner, more. 15 renderings of paintings discussed. 44 calligraphic decorations by Dalí. 96pp. 5⅜ x 8½. (Available in U.S. only.) 0-486-29220-7

ANTIQUE PLAYING CARDS: A Pictorial History, Henry René D'Allemagne. Over 900 elaborate, decorative images from rare playing cards (14th–20th centuries): Bacchus, death, dancing dogs, hunting scenes, royal coats of arms, players cheating, much more. 96pp. 9¼ x 12¼. 0-486-29265-7

MAKING FURNITURE MASTERPIECES: 30 Projects with Measured Drawings, Franklin H. Gottshall. Step-by-step instructions, illustrations for constructing handsome, useful pieces, among them a Sheraton desk, Chippendale chair, Spanish desk, Queen Anne table and a William and Mary dressing mirror. 224pp. 8⅛ x 11¼. 0-486-29338-6

NORTH AMERICAN INDIAN DESIGNS FOR ARTISTS AND CRAFTSPEOPLE, Eva Wilson. Over 360 authentic copyright-free designs adapted from Navajo blankets, Hopi pottery, Sioux buffalo hides, more. Geometrics, symbolic figures, plant and animal motifs, etc. 128pp. 8⅜ x 11. (Not for sale in the United Kingdom.) 0-486-25341-4

THE FOSSIL BOOK: A Record of Prehistoric Life, Patricia V. Rich et al. Profusely illustrated definitive guide covers everything from single-celled organisms and dinosaurs to birds and mammals and the interplay between climate and man. Over 1,500 illustrations. 760pp. 7½ x 10⅛. 0-486-29371-8

VICTORIAN ARCHITECTURAL DETAILS: Designs for Over 700 Stairs, Mantels, Doors, Windows, Cornices, Porches, and Other Decorative Elements, A. J. Bicknell & Company. Everything from dormer windows and piazzas to balconies and gable ornaments. Also includes elevations and floor plans for handsome, private residences and commercial structures. 80pp. 9⅜ x 12¼. 0-486-44015-X

WESTERN ISLAMIC ARCHITECTURE: A Concise Introduction, John D. Hoag. Profusely illustrated critical appraisal compares and contrasts Islamic mosques and palaces–from Spain and Egypt to other areas in the Middle East. 139 illustrations. 128pp. 6 x 9. 0-486-43760-4

CHINESE ARCHITECTURE: A Pictorial History, Liang Ssu-ch'eng. More than 240 rare photographs and drawings depict temples, pagodas, tombs, bridges, and imperial palaces comprising much of China's architectural heritage. 152 halftones, 94 diagrams. 232pp. 10¾ x 9⅞. 0-486-43999-2

THE RENAISSANCE: Studies in Art and Poetry, Walter Pater. One of the most talked-about books of the 19th century, *The Renaissance* combines scholarship and philosophy in an innovative work of cultural criticism that examines the achievements of Botticelli, Leonardo, Michelangelo, and other artists. "The holy writ of beauty."–Oscar Wilde. 160pp. 5⅜ x 8½. 0-486-44025-7

A TREATISE ON PAINTING, Leonardo da Vinci. The great Renaissance artist's practical advice on drawing and painting techniques covers anatomy, perspective, composition, light and shadow, and color. A classic of art instruction, it features 48 drawings by Nicholas Poussin and Leon Battista Alberti. 192pp. 5⅜ x 8½. 0-486-44155-5

THE MIND OF LEONARDO DA VINCI, Edward McCurdy. More than just a biography, this classic study by a distinguished historian draws upon Leonardo's extensive writings to offer numerous demonstrations of the Renaissance master's achievements, not only in sculpture and painting, but also in music, engineering, and even experimental aviation. 384pp. 5⅜ x 8½. 0-486-44142-3

WASHINGTON IRVING'S RIP VAN WINKLE, Illustrated by Arthur Rackham. Lovely prints that established artist as a leading illustrator of the time and forever etched into the popular imagination a classic of Catskill lore. 51 full-color plates. 80pp. 8⅜ x 11. 0-486-44242-X

HENSCHE ON PAINTING, John W. Robichaux. Basic painting philosophy and methodology of a great teacher, as expounded in his famous classes and workshops on Cape Cod. 7 illustrations in color on covers. 80pp. 5⅜ x 8½. 0-486-43728-0

CATALOG OF DOVER BOOKS

LIGHT AND SHADE: A Classic Approach to Three-Dimensional Drawing, Mrs. Mary P. Merrifield. Handy reference clearly demonstrates principles of light and shade by revealing effects of common daylight, sunshine, and candle or artificial light on geometrical solids. 13 plates. 64pp. 5⅜ x 8½. 0-486-44143-1

ASTROLOGY AND ASTRONOMY: A Pictorial Archive of Signs and Symbols, Ernst and Johanna Lehner. Treasure trove of stories, lore, and myth, accompanied by more than 300 rare illustrations of planets, the Milky Way, signs of the zodiac, comets, meteors, and other astronomical phenomena. 192pp. 8⅜ x 11.
0-486-43981-X

JEWELRY MAKING: Techniques for Metal, Tim McCreight. Easy-to-follow instructions and carefully executed illustrations describe tools and techniques, use of gems and enamels, wire inlay, casting, and other topics. 72 line illustrations and diagrams. 176pp. 8¼ x 10⅞. 0-486-44043-5

MAKING BIRDHOUSES: Easy and Advanced Projects, Gladstone Califf. Easy-to-follow instructions include diagrams for everything from a one-room house for bluebirds to a forty-two-room structure for purple martins. 56 plates; 4 figures. 80pp. 8¾ x 6⅜. 0-486-44183-0

LITTLE BOOK OF LOG CABINS: How to Build and Furnish Them, William S. Wicks. Handy how-to manual, with instructions and illustrations for building cabins in the Adirondack style, fireplaces, stairways, furniture, beamed ceilings, and more. 102 line drawings. 96pp. 8¾ x 6⅜. 0-486-44259-4

THE SEASONS OF AMERICA PAST, Eric Sloane. From "sugaring time" and strawberry picking to Indian summer and fall harvest, a whole year's activities described in charming prose and enhanced with 79 of the author's own illustrations. 160pp. 8¼ x 11. 0-486-44220-9

THE METROPOLIS OF TOMORROW, Hugh Ferriss. Generous, prophetic vision of the metropolis of the future, as perceived in 1929. Powerful illustrations of towering structures, wide avenues, and rooftop parks—all features in many of today's modern cities. 59 illustrations. 144pp. 8¼ x 11. 0-486-43727-2

THE PATH TO ROME, Hilaire Belloc. This 1902 memoir abounds in lively vignettes from a vanished time, recounting a pilgrimage on foot across the Alps and Apennines in order to "see all Europe which the Christian Faith has saved." 77 of the author's original line drawings complement his sparkling prose. 272pp. 5⅜ x 8½.
0-486-44001-X

THE HISTORY OF RASSELAS: Prince of Abissinia, Samuel Johnson. Distinguished English writer attacks eighteenth-century optimism and man's unrealistic estimates of what life has to offer. 112pp. 5⅜ x 8½. 0-486-44094-X

A VOYAGE TO ARCTURUS, David Lindsay. A brilliant flight of pure fancy, where wild creatures crowd the fantastic landscape and demented torturers dominate victims with their bizarre mental powers. 272pp. 5⅜ x 8½. 0-486-44198-9